The Secret of U-Boat G-17

A World War II Spy Novel

Dieter W. Gombert

with my Best wishes.
yours
Dieter W. Gombert

Fideli Publishing

ISBN: 978-1-60414-055-2
1-60414-055-0

e-mail: dgombert@tampabay.rr.com

INTRODUCTION

The idea for this novel originated in Baltimore during the summer of 1957. The author was Research and Development Sales Manager for a major US aircraft company at its Nuclear Division, when two German scientists, Professor Hellmuth Walter (see Walter Method) and his Chief Engineer E. Kruska visited Gombert's office.

The Germans offered their wartime invention, know-how and specifications for an unusual propulsion system to our industry. After much discussion they handed a German text and technical description to the author, for further consideration (still in the author's possession).

The discussion focused on the end phase of World War II, when Germany had developed unique and powerful propulsion systems. As a special, proven design, they offered the 'Hydrogen Peroxide Method' or the 'H_2O_2Drive'. This system was the energy basis for the German V-2 rocket and also proved to be an ideal power source for submarines.

The hydrogen peroxide fuel supplied oxygen for the engines and crew. The vessel was able to stay submerged for lengthy periods. During WW2 it constituted a formidable advancement in technology. Shortly, before the end of the war, four of these U-Boats had been built and tested providing astonishing speeds of 25 to 30 knots, at an ocean depth of 180 feet.

Just before the defeat of Germany the existing four boats were scuttled in the Baltic Sea, but one was raised again by Allied Forces, refurbished and successfully tested. Due to the spectacular development of the American "Nautilus" atomic submarine, further research and development for the offered H_2O_2 project raised no more interest and the Aircraft Company declined the German offer.

The following adventurous novel is based on numerous personal encounters, situations and episodes, experienced by the author himself living underground in war torn Nazi-Germany, whereas the following plot is fictional. However, many of the events actually took place in one form or other.

For instance in the First Chapter the "Murder Inc." incident really took place. Also the idea of the Austrian Gauleiter Eigruber to blow up stolen treasures inside the Bad Aussee salt mines, via eight one thousand pound American bombs is available today as printed history.

1944, in a Strasbourg Hotel the author became acquainted with an American OSS agent who was known to have captured an American Sherman tank, very likely as a ploy. Together with an SS-Officer and a French Nazi Collaborator, the American spy initiated this deed while stationed in Paris. It was surely aimed at gaining the confidence of the Nazi bigwigs. The trio escaped Paris and drove the machine into German lines.

Due to special circumstances, see author's book, "Hell was Empty", the author became a suspect for the *Gestapo*. It was during this time when the befriended agent revealed himself and offered his help to have the author escape his Nazi pursuers through American front lines.

The eventual successful escape by the author into Switzerland in fall of 1944 and other circumstances, provided much factual material for this book. Also British Navy reports of a revolutionary Nazi U-Boat, built by Blohm & Voss Shipyard Hamburg Germany and its spectacular breakout through the British Channel in spring of 1945 added further data to the story.

Consequently, a number of events and persons in the ensuing story have been taken from real life. Nonetheless, the overall plot however is entirely fiction.

CHAPTER 1

Early in March 1944, slightly northwest of Kiel, rain showers whipped over the black turbulent bay of the Baltic Sea port Eckernförde, making the gray concrete structure of the U-Boat bunker a forbidding sight. Some small distance away, at an elevated position a group of high-ranking German naval officers and Nazi officials anxiously scanned the choppy bay; while the relentless wind tore at their leather overcoats.

Suddenly, about a mile offshore, the group witnessed a gigantic water eruption. A U-Boat breached the surface at an angle of thirty degrees. Two-thirds of the vessel left the roiling, white water before splashing back to a floating position and settling into a controlled run toward the shore bunker.

It became obvious she had been traveling at an unusually high submerged speed. The heavily over-coated onlookers shouted their amazement by using words like, fantastic and unbelievable. Then after everyone had finally calmed down they left the observation point and walked back slowly along the pier. Some of them still were shaking their heads in disbelief. Eventually, they stepped into the gaping entrance of the bunker while still talking excitedly as the U-Boat slowly approached the same area.

Captain Hilmar Volker waved his arms over his head, after he appeared in the conning tower. He ignored the required Nazi salute, with its outstretched arm, but instead exercised a proud, formal Navy hail with his right hand at his cap. The U-Boat entered the bunker under the guidance of Volker's First Officer; while the group of spectators followed alongside, walking the pier parallel to the moving vessel. The deck's forward hatch cover opened and crewmembers emerged tying the vessel to the pier. The Admiral was the first to approach the now disembarking Captain and after a Heil Hitler salute, the two men shook hands.

Obeying strict naval protocol, the Admiral requested Permission to come aboard with one of his staff officers plus a gray uniformed high-ranking representative of Hitler's 'SS-Corps'. His name was Standarten Führer Heiner Kempe.

"Permission granted, please come aboard"! The Captain said. "I am reporting the unqualified success of our operational test runs in the Hela area off Danzig. Please follow me to the control station below the conning tower."

The three men joined the Captain and entered the boat through the forward hatch opening. The Admiral, an old World-War-One U-Boat veteran himself, was familiar with most of the operational equipment until they reached the aft area, housing the propulsion machinery. Now, the Captain continued his inspection tour commentary:

"As you are aware Admiral, there are four of these newly developed U-Boats available for action. They are all propelled by the newly developed 'Walter' system, utilizing hydrogen peroxide as a fuel and oxygen-provider. This ninety-five percent pure, high-grade chemical is produced at the 'EWM' factory Lauterberg in the Harz Mountains. The plant has the capability of producing 15,000 tons per year. The facility is located deep inside bombproof caverns. There have been considerable difficulties encountered in dealing with the inherent explosive and corrosive characteristic of this chemical; but our scientists have found ways to overcome these problems."

The Admiral nodded his head, while the SS-Officer scrutinized the austere and tight accommodations of the dimly lit U-Boat interior.

The Captain continued, "The U-Boat has a double skinned hull. The outer shell contains a multitude of rectangular openings allowing seawater to Penetrate. Sixteen separate, flexible *Mipolam* plastic bags contain the hydrogen peroxide, or in short H_2O_2. They are interconnected and placed between the spacing of the two hulls."

The Admiral stroked his chin: "Captain, is this revolutionary principle fully battle proven? It sounds like a somewhat uncertain fuel system."

"In its pure form that could be the case, Admiral, but as I said, our scientists have found ways to overcome this danger. The principle is proven as far as possible, under present wartime conditions."

The conversation continued for quite some time, as both men became involved in a number of intricate technical details.

"Captain, tell me about the performance capability of this intriguing vessel."

'Well, our tests in Hela resulted in a submerged speed of 30 knots at 60 meters

(180 feet) water depth. At this point of development we move faster underwater than most naval vessels do on the surface. Adequate oxygen should be available for the crew, allowing the U-Boat to be submerged for much longer periods than any submarine currently in service."

"Also for conventional surface running and conservation of H_2O_2 we have an additional power plant available. We use an economical 210 HP Diesel, offering a cruising speed of 8 knots. Our longest submerged test with the H_2O_2 system, had durations of five and half-hours, at an average speed of twenty knots. Of course we also have a standard 77.5 horse power electric propulsion system we can use to stay submerged without consuming our limited supply of H_2O_2 fuel."

The group went on to the Captain's cabin continuing the conversation for another forty minutes, covering many areas, which were new to the Admiral. Finally he thanked Captain Volker for his most informative boat tour. With a dramatic sounding voice the Admiral stated pompously: "*Standarten Führer* Heiner Kempe of the 'SD', the political branch of the SS-Corps will personally report the success and development of this project to our Führer." Kempe acknowledged this statement with a consenting smile.

Captain Volker scrutinized Kempe unobtrusively, but wisely decided to refrain from any comment.

Patronizingly, the Admiral continued: "I know Adolf Hitler has special plans for these U-Boats, therefore continue training and working with your crewmembers. You'll be getting further instructions from me soon!"

Captain Hilmar Volker loved his German fatherland, but hated Hitler and the Nazis with a consuming passion. He knew the ignorant brown-shirt plague had infected Germany since 1933 and would eventually destroy his cherished homeland. However, his family's tradition, through three generations, had been a commitment to the Navy. Volker was patriotic, intelligent and as a third generation officer, he was professional career navy.

He could not understand how someone as cultured as the Admiral, who had served the proud Navy of the Emperor, could support a Barbarian, like Hitler and his Nazis. The only explanation he could think of was that the Admiral closed his eyes against his better judgment, just in order to continue his very successful carrier and love for the Navy. Volker had heard rumors that the Admiral, during his wartime services, had been wounded in the head, which maybe accounted for his limited grasp of the situation. Volker concluded any sensible-thinking person should be able to see, what would eventually happen to the Third Reich and Germany.

A single civilian stood alone on the pier. The other spectators had left. He was a slender man of almost sixty, wearing a black overcoat and a wide brimmed dark blue hat. A pair of bifocals rested on his serious, almost sad face. His thoughts were in turmoil. Dr. Ernst Rasche, a chemical scientist and physicist from the prestigious *Kaiser Wilhelm Institut* was responsible for the development of the high concentration hydrogen peroxide product. He or "The Professor" as he was called, also loathed the Nazis with a passion' and felt guilty that he had been the prime instrument in developing this powerful chemical for Hitler's war machine. Eight years ago, his wife had been killed in a mysterious car accident, and as she was Jewish, Rasche suspected the Nazis had been responsible; because he had refused the party demand to divorce her on ethnic racial grounds. A so-called Aryan with a Jewish spouse was not permitted to work in the sciences or arts, but Dr. Rasche was needed for the Nazi war machine. Maybe the easiest way to handle the matter would be to kill his wife.

Now Dr. Rasche was waiting for Captain Volker who just disembarked from the submarine, heading toward the Professor.

"It's nice to see you Uncle Ernst", said the Captain, "I thought you might be waiting for me; since we have much to talk about." The title uncle or aunt is used often in Germany, when close friendships exist between families. The children are asked to use this informality, which eventually continues through their adulthood. The men shook hands and walked slowly to the bunkers cafeteria, where they sat down at a quiet corner table, after having picked up their lunch from the counter.

Both came originally from Hamburg where their families had enjoyed a close relationship for years. Volker's father had been a friend of the Professor ever since the time they had studied together at Heidelberg University. Volker's father, also a naval officer, was sent to Heidelberg by the Emperor's navy, before World War One, taking a course in physics.

Recently, Captain Volker lost both his parents, as they became victims in one of the British bomb raids on Hamburg, which turned the city into a flaming inferno. Naturally, the Captain blamed Hitler's war for this tragedy, firmly believing that: "He who lives by the sword will perish by the sword." Hilmar Volker was brought up in a Prussian-Protestant family tradition of Honesty, were a man's word was a solid commitment. Lately, the only conflict he had was the maxim of obeying any order from his superiors, without questioning. After the death of his parents he decided to listen more carefully to the demands of his conscience, telling him that Hitler was a bloodthirsty tyrant, who should be removed from power, either by Germany's military leaders or in the worst case through an Allied defeat of Germany.

The cafeteria was filled with construction workers and navy personnel eating their meager war time rations and talking about their jobs. Nobody dared to speak about the war or its political implications; for one wrong word or remark could land a person in a concentration camp. This caution did not apply to the Professor or the Captain, for they trusted each other and knew they could speak freely at their isolated table, where nobody could overhear their conversation.

'We have to find ways to delay the use of these U-Boats, especially since I believe the war can't last much longer," Dr. Rasche said. "The Allies are going to land somewhere in France soon and the Russians are already well on the way toward Germany."

In spite of his personal feelings, Captain Volker was uneasy to hear this, for he regretfully knew, he had sworn allegiance to Hitler and his war machine. He still remembered the day he was forced to sign a declaration that he would obey and follow all the *Führers* orders. If he hadn't signed the document, he and his family would have been in serious trouble; because the Admiral, a fanatic supporter of the Nazis was known to take drastic measures against career naval officers, who did not pledge themselves to the party and their *Führer.*

"Uncle Ernst, there is not a lot I can do to delay the use of these submarines, I must always keep the safety of my crew in mind," Captain Volker said. "These submariners are a breed apart. They put their lives in the hands of each other and must obey their Commander and they have a devoted trust in me as their Skipper."

"I understand your position," the Professor said. "But you will undoubtedly know how to act when the right moment arrives. As for myself I will do anything to delay the mass-production of hydrogen peroxide. Our work force in the Harz Mountains consists of great numbers of foreign slave labor. Sabotage is a daily occurrence. Of course *Gestapo* agents and a contingent of SD-members are constantly watching us. They are trained to murder even their own grandmothers, if the *Führer* ordered it. However, we will do what we can and what we must, when the opportunity should present itself."

"What worries me most is that this SS-*Standarten Führer* Heiner Kempe, from Berlin's headquarters, is constantly snooping around the U-Boat station. He also belongs to the *Sicherheits Dienst* (Security Service) or in short 'SD' and is under the command of Adolf Eichmann. This group consists of the most inhuman murderers and will use the newly developed U-Boats, probably for some sinister plans; otherwise why would Kempe be interested in our work?"

The Professor had just finished speaking when, *Standarten Führer* Kempe, together with another member of the SS-Corps, entered the cafeteria. Kempe's eyes scanned the tables until he saw Volker with the Professor. Determined, the pair of SS-men walked swiftly over to them.

"Herr Professor, do you remember me? I am *Standarten Führer* Heiner Kempe and this is *Schaarführer* Horst Stolz, my aide and driver. May we join you for a moment? I still have a few questions, mostly involving the travel range of this new U-Boat and its fuel consumption."

The Captain invited the visitors to be seated and noticed that Kempe appeared to be most impressed to meet the inventor of the newly developed H_2O_2 chemical in person.

What Volker and Rasche did not know was, that it was Kempe who was ordered by his SD superiors. to arrange for the assassination of the Professor's Jewish wife. It was Stolz who had faked the "accident" in a most professional way.

Most of the technical description of the chemical and the propulsion system went way over Kempe's head, but he covered his lack of knowledge with enthusiastic praise. He then offered to drive the Professor to his hotel and the Captain to his boarding house.

Stolz was driving when unexpectedly the air raid sirens sounded with their nerve shattering wailing. They parked in front of the Professor's hotel, which housed an air raid shelter. Scanning the sky, they watched the American bombers approaching from the west, seemingly headed towards the city of Kiel, which was a safe distance away.

The weather had improved. Sunlight shone over the unfolding drama and the small park in front of the hotel offered a good view of the sky and approaching bombers. The onlookers guessed the American planes to be at an altitude of some ten thousand feet. About two hundred Liberators approached in tight formation, their unpainted aluminum fuselages glittering in the sky like a school of shiny minnows.

Suddenly a barrage of FLAK shells exploded all around the planes, displaying dark red flames inside black puffs in the air. One of the planes, obviously hit, dropped out of formation and lost control. Bright red flames appeared all over its wingspan. Other planes reacted conducting a routine maneuver, left the onlookers greatly impressed. The bombers immediately closed the formation gap. Suddenly, two white parachutes separated from the falling plane, floating slowly in direction of the observers.

Kempe gave immediate orders to Stolz to drive on and follow the direction of the descending airmen. Within five to ten minutes the parachutes would land in an open field, near the country road they were on. The chauffeur stopped the car and all three men ran swiftly towards the area where they thought the landing spot would be. They arrived just as both airmen touched down some fifty yards apart.

While both Americans struggled to rid themselves of their parachutes, Kempe and Stolz approached them with drawn *Mauser* pistols. Captain Volker knew that these SS-fanatics were ready to execute both airmen on the spot. At that precise moment they noticed letters printed in white paint on the Americans' backs. Across their flight jackets, in five-inch high block letters, were the words: "MURDER, INC."

Impulsively, trying to save their lives, the Captain shouted, "Don't shoot, we'll need these men; because of their jacket inscriptions!"

Kempe hesitated and Stolz looked questioningly at his superior. Kempe, with a diabolic grin on his face, ordered his aide to hold his fire.

"You're right Captain. We'll need these two air gangsters for some special treatment. Let's take them to the city prison."

The city of Eckernförde had a police station where both prisoners were delivered. Kempe immediately telephoned Berlin to report the incident with special emphasis on the words on the flight jackets. On orders from the Propaganda Minister, Josef Goebbels, a camera crew was rapidly dispatched to interview and film these prisoners.

Less than one week later, documentary newsreels were shown all over the movie theaters, presenting the self-incriminating jacket imprints and camera close-ups of these American airmen. Triumphantly, they were shown throughout Germany and all occupied countries.

The Nazi controlled newspapers explained the origin of "MURDER, INC."

as an American gangster group in Brooklyn who commited assassinations for pay. The reaction of the German people was confused. Many thought it to be incredible and certainly alien to German mentality, that a nation engaged in war was capable of allowing soldiers to express their feelings in such a cynical and self-incriminating way. Subconsciously though, many Germans were awed by the degree of freedom expressed by these Americans. No one would ever know what happened to the two airmen after they had served their propaganda purpose; except that *Schaarführer* Stolz was ordered to take them to a Stalag prison camp in Barth-Mecklenburg.

CHAPTER 2

It was one of those stunning April days in Washington, D.C. with white cherry tree blossoms radiating along the Potomac River, decorating the city with a festive atmosphere. However, the warmth and charm of this April day belied the fact that the United States was still involved in a war against Germany and Japan, which had already lasted almost two and a half years. America's huge domestic industry was on a wartime footing, geared to produce enormous amounts of military equipment. The superiority of both the quantity and quality could already be felt in all war-zones, specially the Soviet Union where a steady stream of supplies, via merchant ships, continued to save Russia from collapse.

A single jeep drove slowly with the traffic along Connecticut Avenue in a northeasterly direction. At the wheel was Captain Peter Mertens, a member of the 2nd Ranger Battalion, searching for a certain left-turn side street.

He was excited and curious. Two days before, without warning, he had been hustled away from the battlefield in Italy, where the American 5th Army was bitterly engaged in combat with the tenacious Germany enemy. To his uneasy shock he was ordered without explanation, to pack up and get his ass to Washington immediately.

Eventually, having found the given street, he halted in front of a secluded villa, once owned by one of the enemy nation diplomats.

Mertens was in full uniform including the black Ranger beret. He wore a number of war decorations earned in North Africa and during the almost disastrous Anzio landings. A handsome man, over six feet tall with black wavy hair; his intelligent, friendly blue eyes attracted always-admiring glances from women.

Mertens grew up in Chicago's Rogers Park district, near the dry suburb of Evanston, where he finished high school in 1936. He had a close relationship with his father, who owned a sailboat at Montrose Avenue harbor. He grew up to love sailing on the formidable Lake Michigan. His mother, a French-Canadian, instilled in him a liking for classical music and the fine arts. All of these past events however, had been overshadowed by his recent war experiences and now he was wondering what was awaiting him, here in Washington.

Mertens parked the jeep and went through the high iron fence gates of the villa. Upon entering the building an MP checked his credentials and guided him to one of the many doors on the ground floor. After knocking, the MP opened the door for Mertens to enter, and then returned to his post.

Mertens thought that he had entered a select men's club. The heavy leather armchairs, and tall bookshelves along the walls, in this cigar smoke filled room, confirmed his first impression. A large map of Europe covered one of the walls, with markings and color-pins, depicting the front line in Russia. England also showed many of these pins, especially numerous along the southern coastal areas.

Mertens observed, as he crossed the room, that the floor was of fine polished oak parquet, covered by a huge oriental carpet. Three occupied desks were positioned in front of iron barred windows. An army Colonel rose from the center desk and after salutes were exchanged extended his hand toward his visitor. "I am Colonel Donald O'Connell."

"Welcome to Washington. We have been expecting you. Meet Major Steve McGuire and Captain Richard Johnson. Dick is ex-embassy staff from Berlin. Steve, Dick and I work together. I am heading this section of the Office of Strategic Services. Our OSS task force is involved in obtaining information regarding development of German secret weapons. But, before I continue, take a seat; we first have several things to go over and confirm."

Looking at a folder in front of him, the Colonel stated. "From this I see you were born in Chicago, your father is a German naturalized citizen who came to the US in 1913. He is presently working as an engineer for the New World Oil Company. Correct?"

"Yes, Sirl" Mertens replied.

"Your mother, originally from Quebec Canada, is a concert pianist. In 1936 your parents sent you to Berlin, to attend its Technical University, where you graduated in spring of 1941 as Diploma Engineer."

"Tell me, why were you sent to a German school? What's wrong with our American institutions?"

"My father was of the opinion that the Berlin Technical University was one of the best schools of its kind in the world. The school is closely associated with the outstanding technical brain tank, called the *Kaiser Wilhelm Institut,* a place that produced names like Fritz Haber, Max Planck, Albert Einstein and many other Nobel Price scientists. Also it would give me the opportunity to vastly improve my knowledge of the German language.

Don't forget, I started my studies in 1936, three years before the Germans started the war against Poland and at a time when the whole world admired the Germans during their well-organized Olympics in Berlin."

The Colonel was satisfied with that reply and continued reading from the file folder. "In fall of 1941 you returned to Chicago, fluent in German. After the outbreak of the war you volunteered for the army, where they assigned you to the Ranger Battalion. You participated in the North Africa landings and became part of the operation at Anzio, where the Rangers were almost annihilated. Do you have any comments or corrections regarding these data?"

"No Sir, Colonel!"

Mertens started to examine O'Connell more intensively, seeing a lean seasoned warrior; his face marked by strong weathered creases, undoubtedly acquired on active duty. A half-smoked cigar stuck aggressively from the comer of the colonel's mouth. His deep set flashing eyes signaled no acceptance of disagreement, unless the reply was well conceived and prepared. There was, however, around those battle weary eyes a hint of sardonic humor.

From behind the desk and through the blue haze of smoke the colonel said: "Captain, we have come to the conclusion you are ideally suited for this most sensitive operation we have in mind. You're an engineer, you speak the right languages and for this task we need someone with your technical training, capable of analyzing and interpreting enemy data. You must be willing to go behind German lines and become active as an agent. You realize this means you must live and move in civilian clothing, through occupied territories. It also means you get shot as a spy, if you get caught."

At this point Captain Johnson, a Scandinavian blond fellow, interjected by speaking German, which he spoke fairly well, but with a heavy American accent. "Do you believe your German and French will be good enough, while you are amongst the enemy? You know your life will depend on it. At Langley we'll check you on this matter; to be sure you won't get into trouble."

Mertens replied in flawless German, which even had a slight Berlin accent: "I think I speak German like a native and my French, learned from my mother, is pretty good too.

I only have to watch sometimes, to avoid a little bit of a Canadian inflection."

Being pleased with Mertens' reply and confirming Johnson's statement, the Colonel continued: "The object of this particular operation is to make contact with a German scientist, who had escaped the *Gestapo* and who is presently hidden in the formerly French province of Alsace. This Professor knows of Germany's latest secret weapon programs and he is prepared to hand over vital information, provided we can get him and his adult daughter out of Alsace. Arrangements should be made for father and daughter to escape into Switzerland. Of course, the Professor's vital information would enable us to take necessary counter measures. Would you be prepared to take on such an assignment?"

Mertens thought and remembered the good times he had enjoyed in Germany and Europe, before America entered the war. Echoes of laughter, friends, fun and love, flooded his memory. But he also recalled an incident in 1941, shortly before he returned to the USA; when his Jewish landlord with his family was brutally arrested and deported to the east. He remembered something he had heard recently and it bothered him. It went something like: 'Evil will succeed where good men stand by and do nothing.' "Yes sir, it sounds challenging. Count me in!" Mertens said eagerly.

"Excellent, Captain! Let's get to work immediately. Forget formalities, first names will do. Welcome to the Office of Strategic Services. There's a room upstairs where you can live and work after your briefing. We'll notify your Ranger unit of your transfer.

"Now listen to this: Through the French underground our contact has advised, that the Nazis have developed a new chemical propellant, and are producing special engines for this medium. It is said the application would be useful for rockets and U-Boats, but we have neither further details nor information. It is vital to evaluate the danger stemming from these inventions. Any information would be of interest to our counter measure sector!"

Receiving a nod from the Colonel, Major McGuire continued the conversation. "The Allied Forces will soon undertake gigantic landings in Europe, specifically through France. Immediately, after our forces. are solid on the ground, we want you to penetrate German lines, where your goal will be to reach Strasbourg, which the Germans have annexed from the French, as part of Germany. Once there, your contact will be a Prof. Dr. Ernst Rasche who lives hidden in an old part of town with his young daughter.

Now, Mertens became excited. He had heard of Dr. Rasche, while at the Berlin University. He remembered some of his own 'visits at the *Kaiser Wilhelm Institut*, where he saw the name, Prof. Dr. Ernst Rasche, posted at one of the office doors. He recalled the story of the Professor's wife being killed in a mysterious automobile accident. Since then, the Professor was supposed to live alone with his beautiful daughter.

Still musing about these facts he suddenly heard Major McGuire's voice again. "Time is at a premium, we will have to compress the preparations for your assignment to a period of one month. Our people in Langley will provide further upgraded information to you. In the meantime while you are here, use all the facilities available. Keep the Langley jeep, since you obtained it from them in the first place when you arrived from Europe. I'll fly to England shortly where the necessary preparations and arrangements for your assignment will be made. Dick Johnson will leave for Switzerland as soon as the Allied landings have proved to be successful. Dick will be stationed at our Zürich Consulate, to act as communications center. He will be your contact with the Professor and he'll stay in touch with Washington, through consulate cipher. A planned underground radio station in the Vosges-Mountains should be created and handle everything covert and local.

"The operation is of course top-secret. We thought the password "Blondie" appropriate, as we believe the new propellant is a concoction of hydrogen peroxide. You may inform family and friends that you are based in Washington, but of course, not a word concerning this project shall ever reach the outside!" Finally, Colonel O'Connell concluded the meeting by saying: "Once McGuire has left for Europe you can take over his desk and we'll be able to work together. As we said before, the preparation time has become dangerously short. We'll have to make extensive efforts to coincide with the proposed landing date. As it would appear this German scientist is a genuine defector, with a wealth of high level technical information. We can't afford to lose him.

"A daily program has to be developed, indicating where and how you can make contact from France and Germany, to reach Dick in Zürich. Also our French underground agents, I am sure, will be of great help.

"We'll have to create good quality identification documents for you, including current food stamps, proper currencies, local clothing, supplies and so forth. Langley will play a major part in the acquisition of these logistics. Welcome aboard Peter! We are pleased to have you with us!"

CHAPTER 3

June 12, 1944, heavy waves of the English Channel hammered relentlessly against the steel-hulls of the many landing crafts. The occupants of Mertens' landing vessel had set out many hours earlier, from Dartmouth in the south of England, aboard a tank landing craft. Later they changed to the smaller troop vessel, used for landing personnel in shallow water.

The monotonous droning of the engine propelled Mertens' boat toward Normandy and Utah beach. He and Steve McGuire were half sitting, half leaning on one of the munitions boxes; from where they apprised the group of thirty soldiers of the 4th Infantry Division. Most of them were standing, and all looked dejected and decidedly unwell from sea-malaise. The soldiers tried in any way they could, to survive the pounding of the heavy seas. Once in a while a shower of salt water sprayed over every inch of the interior. It was 2100 hours with darkness soon to cover the beaches, as the sun descended slowly, beyond the ocean's horizon.

The day had been sunny and warm, although during the morning hours a light fog had hovered over the sea. Many of the soldiers had been seasick and the gut-wrenching odor hung in the air. Nobody spoke. Everyone was occupied with thoughts, how well would he fight, would he get wounded, would he survive?

Peter Mertens' reflection wandered back in time when he lived in Rogers Park, Chicago with his parents. He was their only child. They owned one of these typical frame work bungalows, with a wide front porch, shaded by huge elm trees. It wasn't far from their home to Lake Michigan, which could easily be reached, in a fifteen-minute walk. Even on the hottest summer days the lake was terribly cold and temperatures rarely exceeded seventy degrees Fahrenheit.

Mertens loved the water and the vastness of the sea, sailing on Lake Michigan was not always an easy undertaking and he remembered moments when he and his father were caught in nasty squalls, which seemed to appear quite suddenly, out from no-where. Therefore, he sort of enjoyed this trip toward Normandy beach. It was only his upcoming assignment, which left him with anxiety and hope, that he would be successful, and most of all luck would be on his side.

On June 6th, the first Normandy landings had taken place and within three hours of hitting the beach, US troops had penetrated four miles inland liberating the town of Ste.Mere-Eglise. The night before, US paratroopers had softened up the occupied town, catching the German defenses by surprise.

At the time of the landings no German General dared to take decisive action, without receiving instructions from Hitler, or Gröhlfatz as he was called by his disgruntled officers, which is the German abbreviation for. "Greatest living Field Marshal of all times". Gröhl in English means 'bowling, and Fatz or Fatzke stands for, 'ridiculous clown'. Hitler was having his morning nap and nobody was allowed to wake him, prior to ten o'clock.

The Americans had taken Carentan, about 12 kilometers south of Ste. Mere-Eglise, but British Forces were still bogged down around Caen.

More and more noticeable, above the sound of the landing craft's engine, was the rumbling noise of artillery fire, together with the rattling of machine guns. The craft was getting closer to the beach, by-passing several Allied war ships, which were firing their deadly, high explosive shells into Normandy. The noise of each naval-vessel's huge gun detonations was earsplitting and the reverberations were intensified, as they traveled across the water.

Mertens was dressed as a French peasant, wearing a dark blue beret and carrying the necessary identification documents, food-stamps plus a few hundred Franc notes in his pockets.

He hadn't shaved the last few days, the black chin-stubble giving him the appearance of a native. He wore a pair of old worn leather boots and the only weapon he carried was a folding switchblade knife.

The sound of the engine changed and the landing craft slowed down. Mertens and Steve climbed on the boxes and viewed the approaching beach of light colored yellow sand. Beyond the beach, the flashing fires of exploding grenades and the illumination from flares brightened the horizon. The crunching of the hull over the beach, alerted everyone into action. The bow gate dropped into the water's shoreline and the soldiers started to disembark; wading chest deep toward the shore, carrying their weapons high above their heads.

As previously arranged, Mertens was carried ashore by three soldiers to keep him from getting wet. If the Germans should capture him, salty clothing would reveal that he had arrived by sea. Steve McGuire knew what he was doing, but had to wade ashore.

In the twilight, a long line of stretchers appeared, accompanied by a group of medical corpsmen, waiting to bring wounded soldiers on board the emptying vessel, for their return to demurring hospital ships. A crew of GIs labored on a concrete landing pier, unloading various supplies, which were strewn and piled, untidily all over the beach. Debris was everywhere, empty crates, discarded K-Rations, broken equipment, empty cartridges and every type of war waste imaginable. Obeying their NCOs and officers, hundreds of soldiers were organizing themselves into various groups, ready to reinforce the front lines.

Steve McGuire went on ahead, in direction of the sand dunes, with Mertens following until a narrow winding path into the dunes became visible. Two paratroopers were standing there, one holding an old bicycle, while the other carried a worn and soiled knapsack on his arm. After a short exchange of salutes the Major gave the password: "Blondie!" The immediate answer was: "Jean Harlowl" Identification was correct; the contact had been made.

One of the paratroopers handed over the bicycle to Mertens, the other the knapsack, which contained cotton shirts, underwear, socks, shaving gear and a small supply of food; all of which were obtained by swapping other goods with local farmers.

From their elevated position the wide Atlantic Ocean was still visible in the fading daylight presenting an overwhelming spectacle of hundreds of different size ships and landing craft, while directly below them, on the beach, soldiers teamed up everywhere.

"What is the latest report on our advance?" asked the Major.

"Pretty exasperating!" replied one of the paratroopers. "The Krauts are dug-in at Montebourg, putting up a hell of a fight. As a result Cherbourg harbor is blocked. Also the areas around Bayeux and St.Lo are being strongly defended. The Brits have a hard time getting through!"

"The only language the Krauts will understand is a well placed, concentrated air-attack!" replied Peter, remembering his stint in Italy.

After walking half a kilometer along the dunes, the four men reached a resort hotel, which once served vacationers from Paris. As a result of recent fighting, the roof and the left wing were heavily damaged, but a group of soldiers were busy doing repairs.

Clearing his throat the Major explained. "This place will be a good command post from which we can organize, coordinate and supply our actions. Mertens, you'll leave from here with a patrol going behind the German lines about 0100 hours, just before the moon rises. Squad Sergeant John Broden will lead the patrol; his experience from Salerno and Monte Cassino will be invaluable. We also assigned a few paratroopers to the patrol, who have seen action and who have already been behind German lines. These men, two of whom you have just met, should be of great value. They are familiar with the terrain. Finally, the bicycle should be useful in bringing you, undiscovered, to your first assigned meeting place in St.Cloud."

After having passed two MPs at the hotel entrance, the four men stepped into the hallway where they were met with a hive of activity, in spite of the late hour. Members of the signal corps were busy laying and connecting various cables, while others made the place functional, including repairing damaged areas. Several officers of the Eighth Infantry Regiment were shouting orders over newly rigged phone lines.

A young Captain greeted the Major and, following a short introduction, they were led into one of the adjoining rooms, where a group of soldiers were leaning over a table, covered by a military map.

"The Blondie group just arrived!" the young officer announced to the gathering consisting of an army Sergeant and six paratroopers. The men stood at attention when the senior NCO saluted and said, "I am John Broden!"

"Good to meet you Sergeant!" replied the Major. "I presume you are familiarizing yourselves with tonight's patrol orders? What does it look like? Have you found a safe way behind the German lines?"

Sergeant Broden referred to a paratrooper Corporal who answered. "We believe northwest of Marigny is the most favorable point to slip by the Krauts. The area is lined with these lousy hedges, but in this case they'll definitely be to our advantage. I estimate a distance from here to St.Lo, where our forces are held up, to be about 30 kilometers. The town is currently defended by German elite-troops, namely their 3rd Parachute division. St.Lo will be a tough nut to crack, but on the brighter side, the first 25 kilometers to our destination can be driven by truck, and from that point to Marigny should only be about eight more kilometers or five miles."

Mertens, now dressed as a French farmer, leaned over the map in an effort to familiarize himself with details. After memorizing as much as he could, he turned to Steve McGuire and said, “I'll find myself a barn around Marigny and rest till morning. Civilians are not permitted to be out at night, certainly not on the roads.”

“Correct”, said Steve. “Why not find a spare room and catch a couple of hours sleep before you have to leave?” Mertens followed his advice, feeling weary after a long and strenuous day.

At midnight Mertens awoke from a short but deep sleep as Steve shook him back to reality. “Your men are outside,” Steve said. Mertens was immediately awake and alert, for as a combat soldier he had learned to seize the opportunity to sleep, whenever and wherever possible. He knew that a tired soldier could easily panic and turn a minor difficulty into a major crisis and make bad mistakes of judgment.

Mertens picked up his gear and went outside where the patrol squad of nine was already seated in an open truck, each with blackened faces, a sub machine gun hanging from their necks and spare magazines strapped all over them. In addition, John Broden wore a pair of night-glasses.

As he left, Mertens shook hands with his new friend, Steve McGuire.

“Good luck. Keep your head down!” Steve called out to him. Peter put his bicycle onto the truck, got on board and the group took off into the night.

With dimmed lights the truck followed a narrow country road, bordered on each side by a barrier of tall hedges, preventing any sort of clear view over the countryside. They were not alone on the road. On their right marched a seemingly endless column of GIs heading toward the front lines in single file.

After half an hour's ride, the truck had to slow down due to a number of Sherman tanks moving sluggishly ahead. Finally Mertens' truck was able to regain speed and direction. Using a split in the road, the tanks moved left, while the members of the patrol chose the road to the right.

About another half-hour later the truck came to a stop. “That's all folks! We don't go any further!” said the driver.

He had halted the truck a short distance from a larger road coming from the West Coast of Normandy. At a command from Sergeant Broden the paratroopers jumped from the vehicle and spread themselves out, crouching along the hedges.

The truck reversed into a road cut, turned, and disappeared into the night.

Almost immediately as if on cue, a young Lieutenant and his platoon Sergeant emerged from the darkness. After recognizing Broden, the Lieutenant welcomed them. "Glad you made it, no fire here at this time," he said. 'We were advised about your coming only about an hour ago. My men are dug in about one hundred yards south of here, on our part of the road leading to St.Lo. Close to St.Lo the Krauts are thick on the ground, hidden behind hedges, along their part of the road, which at night is very active. During daylight they don't use any of the roads much, because of the targets they would make for our Thunderbolts."

Apart from the continuing distant sound of rolling, moving enemy tanks the front-line at this point was astonishingly quiet, although once in a while a bright flare illuminated the landscape. The Allies seemed to be waiting for the next day to continue the war, probably waiting in order to accumulate more men and equipment, to eventually break through to St. Lo.

Mertens, the Lieutenant and the two Sergeants stood near a ditch reviewing an area map with flashlights; showing the German positions clearly marked in pencil, when a Corporal appeared from the darkness. "Lieutenant, we just got a message," he said. ''You are to call headquarters immediately; they seem to have special instructions for you!"

The Lieutenant, together with Mertens and John Broden made their way to the outpost dugout, situated in between the hedgerow. The Lieutenant picked up the field phone and said. "Caniglio here, Bravo section. What's the problem?"

''Wait.... Colonel Menken wants to speak to you!"

After a short pause the Colonel was on the line. "Lieutenant the 'Blondie' unit is with you, correct?"

''Yes Sirl" answered the Lieutenant.

"Very well! I want you to immediately arrange for a patrol of fifteen men to penetrate enemy lines. The Krauts are planning something and we need a few prisoners to find out what they are up to. Heavy traffic moving over the road to St.Lo gives us an uneasy feeling, and the patrol will be an excellent opportunity to put 'Blondie' in position to move cross-country. Put Broden in charge, utilize paratroopers and make up the necessary numbers from your forward group. Naturally, a prisoner 'snatch patrol' and the placing of Blondie must be accomplished before daylight. Understood Lieutenant?"

"Understood Colonel! We will move immediately!"

The time was about 0130, the weather pleasant but cool. Behind the enemy lines, a half-moon appeared slowly from the horizon, rising into a gunmetal sky, filled with a million stars, illuminating the landscape. The Lieutenant selected six of his best men and transferred them to Sergeant Broden, making a patrol of fifteen soldiers plus Mertens. Broden moved into a forward position, where lying on his stomach, he scanned the enemy lines and the road to St.Lo, with his night binoculars.

He decided to wait a while, maybe the Kraut night traffic would soon abate. Scanning around, he saw an open field about 150 yards wide, lying between his group and the enemy. Further to the right he noticed a low fieldstone wall, along which lay his best chance to advance unnoticed.

The patrol waited patiently for about an hour, after which, as Broden had hoped, the road traffic thinned down. At a given signal from the Sergeant, the men were up and crouch running, at fifty-yard interval spacing, toward the wall. 'Mertens was the fifth to run in the group, unfortunately he had to abandon his bicycle during the movement; however his knapsack was securely strapped to his back. All sixteen men crossed successfully, even as an enemy flare descended from the sky, its savage white light searching out targets for destruction. The wall however shielded them sufficiently from detection and after the flare died, darkness temporarily returned. Swiftly the men arose running, heading again along the wall, east into enemy territory. They stopped a second time for a flare, but the wall still protected them. Convenient bushes covered their advance and the enemy was still unaware of the approaching danger.

During the next dark period, at a given signal from Broden, the men lunged forward over the open field, approaching the fifty yard distant hedgerow, emptying their sub-machine guns into the bushes at the hidden foe. Twelve to fifteen hand grenades placed into the hedge, clearing it of opposition and ripping a hole through which the group could pass. The attack was a complete surprise and success.

The German defense at this point was undermanned. They put up only token resistance, firing just a few rifle shots. Within a short time the patrol was on the highway, where they turned left and right, and boring back into the enemy's flanks. Mertens also on the highway, grabbed this opportunity, raced from the road into the adjacent field, then moved steadily in a southeast direction. He was quickly surrounded by darkness and relative safety. The fifteen men he just left were engaged in the exchange of small arms fire, now a distance away from Mertens.

Walking through thick grass in semi-darkness, Mertens saw another hedgerow, about fifty yards to his right. Approaching it he decided to stop and rest under cover of the hedge, which grew along a wall of earth. He noticed at this point, that there was a narrow road on the other side of the hedge. Having worked his way into the brush, Mertens settled down and listened to the night. The uneasy silence was eventually broken by a fire fight between lighter sub-machine guns and heavier German weapons; the conflagration finally ending in a few thudding hand grenade explosions.

It was quiet when Mertens decided to get on the move again. He began to walk carefully on the narrow path, still heading southeast. He recalled the map of the area, which he had memorized back at the base and realized Marigny could be only about five kilometers away. Suddenly, he heard human noises and the rhythmic clanking of backpacks and side arms, moving toward him. Instinctively, he was on the ground rolling into the bushes for cover.

Almost immediately, not ten yards from him, a company of sweating, heavily laden German soldiers jogged past in a northerly direction. He thought these enemy soldiers were undoubtedly a backup, called to strengthen the weak spot on the road. The penetration of the American patrol must have caused counter measures.

According to the old pocket-watch, which Mertens carried in his soiled jacket, it was 0300 hours. The dangerous situation had passed and again he headed on toward Marigny, which by his reckoning he should have reached by now. To get some idea of his location, Mertens broke through the hedge and saw in the half-light of the moon, a large barn some distance from an even larger farmhouse. He assumed all would be occupied by Germans. The curfew for civilians was still in force until 0700. Realizing it was essential to have enough sleep; Mertens took advantage of a grassy cavity in the hedge, not visible from the road, to catch a few hours rest. Crawling into the hideout and using his knapsack as pillow he fell asleep almost instantly.

An enormous thundering awoke him at about 0700 together with bright sunlight in the sky. Waves of planes were droning overhead. Looking through the hedge, he could see black smoke clouds billowing into the air, some several miles distant, obstructing the morning brightness. The planes unloaded their deadly cargo onto a target, which Mertens assumed to be St.Lo. Finally, the planes headed west out to sea. "Was this the beginning of the Allied attack against the city?" he thought.

Mertens started to reassemble his memory. "Marigny is not the best area to be heading for; it's high time to get out of here, away from the front line."

Further to the south he picked out another country road. He knew it would be the highway coming from the near coastal town, Coutances, in western Normandy and leading east into St.Lo.

He made his way toward the road and almost made it when a Wehrrnacht motorcycle with sidecar sped by, traveling in a westerly direction. Crossing over onto the southern edge of the road, leaving Marigny to the east, he began walking west ten more kilometers. The old pocket watch gave the time of ten in the morning. Traffic on the road was becoming more and more active, particularly the number of ambulances chasing in either direction. On several occasions Mertens had to run from the side of the road, seeking shelter in a ditch and seeking cover, to be protected from strafing air attacks. These marvelous Thunderbolts knew their business, but fortunately he was never in direct danger.

Not having eaten for many hours and his provisions being used up, hunger was becoming a problem. Mertens decided to continue another five kilometers to the town of Coutances and away from the immediate action. It was 12:00 noon when he eventually arrived in the town's main street and discovered a small bistro.

He entered the restaurant and sat at one of the uncovered wooden tables. A bar with a brass foot-rail stretched along the back wall. The establishment's shabby furniture added to the room's austere appearance. The proprietor, a man of about fifty years, his chin covered with bristly unshaven stubble, walked over to his table. His collarless open shirt was decorated with a single brass-collar button.

"Que peux-je vous offrir?" he said, or "what can I offer you?"

"I have food coupons and I'm very hungry," Peter replied in French.

After an unidentifiable bowl of soup, the innkeeper served roasted potatoes with a small pork chop, red beets and endive salad. The salad was moistened with a dressing of watered vinegar. A glass of Beaujolais arrived as an unexpected bonus.

'Where do you come from?" asked the host.

"I originate from LaGorgue near Lille. I was visiting my sister on her birthday, in nearby Valogne, when the invasion forced me to get home as fast as possible."

These words started an emotional conversation, the main topic of course being the invasion. First the innkeeper scanned the room, to make doubly sure they were alone, with no German soldier present. Then he excitedly gave his expert account of the invasion. Enthusiastically, with tears in his eyes, he raved about this fantastic invasion.

"Finally we have a chance to chase this hated *'Salle Boche'* from our French soil," he said. Then he explained in detail about all the places where the Americans landed. "By now they have surely taken Cherbourg and in one week they'll be in Paris!"

Sitting near the window gazing into the street, Mertens noticed a country Gendarme arriving at the bistro on a bicycle, which he carefully pushed inside the designated parking stand. He wore a dark blue uniform with a typical French cylindrical cap, and a leather holster containing a pistol, a wide leather belt and shoulder strap supporting his gear.

"Be careful, this swine is a Nazi Collaborator, comes here every day to eat his own sandwiches, while drinking my precious red wine!" The innkeeper hastily warned Mertens.

"*Salut*!" said the policeman after he had entered, staring suspiciously at Mertens. "*A oui*, a stranger in our town, *n'es pas*. Where do you come from?"

Peter repeated the story about his fictitious sister, after which the civil official made a facial grimace, not due however to the story, as Mertens first feared. The gendarme suddenly disappeared in the direction of the toilets.

Mertens recognizing this great opportunity, quickly paid for his meal, nodded to the innkeeper, gave him a knowing wink and in a flash was on the street. The bicycle was exactly what he needed. Seizing the unlocked bicycle and pedaling furiously, he rapidly left the area. The innkeeper in due course, put on a convincing performance of being completely distraught and rattled by this audacious theft, calling: "*Monsieur gendarme, Voleur! Thief! Voleur*!" Alarmed the gendarme appeared in his long white underwear, trying to pull his pants up and looking absolutely ridiculous.

In no time at all Mertens had left the little town some distance behind. Pedaling back to where he came from, he remembered from his walk into town, that there was a field path leading in a southerly direction, finally disappearing in the distance over a hill. This little-used, sandy path was exactly what he needed, provided he could find it again. Luckily he spotted the trail. Cutting across a field, he joined it and within just over two hours he was in Avranches.

Approaching the town, Peter noticed heavy traffic going north on a parallel highway. The German *Wehrmacht* was desperately trying to strengthen a counter-offensive. Green tree branches trying to hide themselves from the enemy air force covered their vehicles. In spite of this camouflage, attacking deadly Thunderbolts, constantly harassed them. Several times Mertens had to scramble for his life, escaping this 'friendly fire.'

Bypassing the town of Avranches he continued his journey south over Fougeres to Vitre, where he discovered a farm estate. Since there were no Germans in the area, he decided to talk to the farmer and his wife, an elderly couple who turned out to be very hospitable. Mertens had to report his latest impressions of the fantastic invasion, bringing the couple to tears, when he reported all the good news. They invited him to a sumptuous meal, which actually turned into a celebration.

Sadly, the couple told of their son being taken away to Germany, where he was forced to work for a farmer; while the chores on their own farm, without their son, were breaking their strength. Mertens believed he could trust his newly acquired friends and retreated to a refreshing, long and deep sleep in the barn.

The next day, after an emotional farewell, he bicycled 110 kilometers to LeMans were he arrived at about 5 p.m. After last night's agreeable experience, he decided to move on, and find another friendly farmer. The hard saddle was causing him considerable pain and discomfort, so he was looking forward to a rest. Some 10 kilometers outside LeMans he was lucky to find another farm, where he was again allowed to sleep in the barn.

The following day he passed the city of Chartres and kept on going through Versailles. In the afternoon, he reached the first leg of his destination, the Paris suburb of St.Cloud. The town spread over a high bluff, overlooking the river Seine and panorama of Paris. After a few inquiries he found the house of his designated contact in the rue des Binelles, an old gray facade structure with three stories, and windows covered by gray wooden shutters. Three worn out stone-steps led to the heavy oak door entrance, where a brass plate announced the profession of the occupant: "Avocat", or Attorney.

Leaning the bike against the wall Peter rang the bell, waited a short while, until eventually a middle aged Frenchman opened the door. The man had expressive light blue eyes, a round face with rosy complexion, and sparse blond hair. Standing in the entrance the man said: *"Bonjour Monsieur!"*

Mertens replied: "Could I talk to Blondie?"

Looking him in the eyes, the man said: "You mean Jean Harlow?" With a big smile both men shook hands and embraced. "My name is Bernard! Please come in, and bring your bike into the hallway! I have been expecting you!"

The entrance hall was dark, but still light enough to reveal a door on the right, which led inside to a modest office. An open roll-top desk took up the entire opposite wall space. The other walls were covered by bookshelves containing law reference books, together with cardboard ring binders. In the middle of the room stood a wooden table with four chairs. An ancient black dial telephone rested on the desk, and two windows gave some skimpy light, through semi-closed shutters. Bernard switched on the glass plate ceiling lights, illuminating a couple of leather arm chairs, a small round table and a floor lamp.

At Bernard's suggestion both men sat down. The host opened a bottle of Napoleon Cognac and poured drinks into two bulbous glass tumblers. "Tell me the latest news of the invasion," said Bernard.

"Well, I believe we have the *Choucroutes,* (Sauerkrauts) on the run. It is only a matter of time and the Allied Forces will break out of the Normandy area, heading directly for Paris. We are having trouble liberating St.Lo, however I believe a full-out offensive has just begun, and they're bombing the hell out of that town. Our troops are bogged down toward Cherbourg and the Brits are behind schedule in taking Caen."

"As you probably know," Bernard said. "I am working for Steve McGuire with whom our French underground system has had steady contact for about a year. Yesterday I received a message from London relating to your assignment. The liberation of Dr. Rasche in Strasbourg has been put on the back burner. It appears that Allied Forces will soon be in Strasbourg anyway for his liberation."

"London has gotten word that a tough, high ranking Nazi officer is staying at the Hotel 'de Crillon' in Paris. He is SS-Standarten-Führer Heiner Kempe, one of Himmler's favored henchmen. Through Berlin they have heard, he is to be, or may be involved in the use of some secret submarines, we are not quite sure which. They also suspect the Nazi hierarchy is planning something big and we are very concerned about it. Your new orders are to stay at the 'de Crillon' and get to know Kempe.

Peter looked at Bernard with a surprised expression, "Did you say Heiner Kempe?"

Bernard replied: "That's correct, why do you ask?"

"I knew of a Heiner Kempe, while I was a student in Bertin. At that time I rented a room from a Jewish family, who treated me like their own son. They had a lovely daughter, my age, whom I started to take a fancy to. One morning in February 1941 at about 6 a.m. we heard some violent knocking at our apartment. My host Herr Siegfried Kohn opened the door. Then, rushing into the apartment were two Gestapo agents, accompanied by two SS-Men in uniform. They rounded up the Jewish family in a most brutal manner. With shouted orders they forced the family to get dressed in a hurry. Then they entered my room and one of the SS-Men, carrying a baton under his arm, asked me for my identification. After he discovered I was an American he clicked his heels, excused himself and told me he was under orders from Standarten-Führer Heiner Kempe, to evacuate the Jewish family to a better location in the east.

After Herr and Frau Kohn, with their daughter Miriam, were dressed they tried to take some of their belongings with them; however, this wish was shouted down and with rifle butts the SS-Men drove the family out of the apartment, leaving me alone in the place. I ran to the window and saw parked below, at the curbside, an open military truck. The back gate was open and the family was shoved and lifted into the truck. Then, to my utter horror and after the gate had been closed, I observed the SS-Men flinging a heavy rope netting over the platform opening and fastening the trapping device to the sides of the vehicle. The family was caged like a bunch of chickens being brought to market. Then the truck drove off into the darkness of the wintry morning. I swore never to forget this event and the name Heiner Kempe will always be fixed in my mind. Also that son of a bitch SS-Man, with the baton, I shall not forget. Unfortunately his face was not too clear in my memory since my room was fairly dark at this early hour of the day. What a weird coincidence to have me become acquainted with this bastard Heiner Kempe."

Bernard was most agitated by this story and he was amazed at the turn of events. Then he said, "I hope you won't meet this SS-Man with the baton again, he could possibly recognize you."

Peter answered, "I don't think he would remember my face because of the room's darkness. I only hope he won't remember my name, should I ever run into him again."

Bernard continued his explanations. "I am a widower and this is my home as well as my law office. One of our agents, who doesn't live far from here, owns two donkeys with a small wagon, used for supplying straw to his goat and rabbits. Hidden under the straw is a radio-transmitter, which enables us to keep in contact with everyone; while our man moves his fodder. My job is to receive and coordinate all pertinent intelligence from our sectors, and keep London up to date. So if you have any traffic to send, come to me.

Now, while you are with me you'll receive a German travel passport, a military passport and a document with picture, proving you are a private detective from Berlin. Furthermore, you'll be provided with a letter from an art dealer, stating you have been hired to obtain and handle valuable art treasures, which have to be transferred to a safe haven in Germany. Of course, you'll also receive a few thousand francs and food-coupons from me. All I need from you are three passport pictures, which a neighbor photographer will immediately prepare for us."

"In the rue du Faubourg St.Honoré you'll find an art dealer at an old antique shop, the owner is René Martin who works for us and will be your cover. He can be used as contact for telephone messages. Since you'll be residing at the hotel 'de Crillon' on the Place de la Concorde, you can't continue wearing your peasant clothes. René will give you a complete set of suits and clothing, made in Germany, together with a necessary suitcase. I'll call René immediately and arrange for him to be at his store tomorrow about ten a.m., you can use your bike to get there. Tonight however, after having your passport photos taken, we'll go out to my favorite bistro and indulge in a hearty meal. The owner is a dear friend of mine and he'll take personal care of us."

CHAPTER 4

Late in the morning of June 16th, on his bicycle, Mertens crossed the bridge over the river Seine. The spring weather was unusually warm, with a blue sky smiling over Paris. At the *boulevard de la Republique*, he turned right moving until he crossed the Seine again. The traffic was light. Most Parisians used bicycles. Only collaborators had the privilege of driving automobiles, readily recognizable in their black Citroen sedans with the manufacturer's chevron markings on the radiator. Mertens continued on the East Side of the river, passing the imposing Eiffel-Tower on the right. Beyond the *Quai d'Orsay* he made a left turn over the bridge to the *Place de la Concorde*. Mertens had to chuckle when he viewed the obelisk. He remembered the story when Napoleon was in Egypt and his wife Josephine had a lover in her boudoir. An assassin penetrated into her room and believing Josephine was sleeping with Napoleon murdered the lover instead. In typical French nonchalance, it was decided to commemorate the victim with the Egyptian obelisk to symbolize sexual manhood.

'What a beautiful city!" he thought; everything so marvelously planned, every tree placed to enhance the appearance of the wide boulevards. Often the scene became distorted with the appearance of *Wehrmacht* vehicles and German soldiers roaming the streets.

After passing the awe-inspiring Concorde with its artistic fountain and historic obelisk, he entered rue Royal, where at the end of the street he saw the Greek, and temple-like 'Madeleine' church. One block prior, he reached *rue du Faubourg St.Honoré*. There he turned and walked a couple of blocks, until he found the antique store he was looking for.

'Gallerie Martin' was spelled out in golden letters above the store, which could only be reached by passing through a narrow arcade recess, leading up to the entrance door. The windows had a display of numerous artifacts and knick-knacks, including a bronze-cast boy, holding a large seashell lamp above his head. A couple of easel supported paintings of the romantic period, showing scenes along the banks of the Seine. On a shelf, in the store front window, lay an assortment of

sabers, old dueling pistols, daggers, and front-loading rifles.

A mellow toned bell sounded after Mertens opened the door. Inside, draped over some stands, was an assortment of Persian rugs. From the ceiling hung a couple of crystal chandeliers, and the walls were covered with numerous oil paintings and gobelins. Standing on a four-foot high marble column was a bronze bust of Napoleon Bonaparte. Displayed on a separate table were selections of old leather bound books. A glass case standing alongside contained all kinds of ancient jewelry. At the back he noticed a red velvet curtain leading to a passage from which a middle-aged man appeared.

"Bon jour! I would like to talk to Blondie!" Peter said in French.

"Oh, you mean Jean Harlow! Yes, I have been expecting you! Bernard told me of your arrival!" Both men shook hands.

René Martin was a slender, small man of about forty-five, with black hair combed back flat. His face had the typical southern French, olive tan complexion, with dark brown observant eyes, revealing a person who had seen the world. He wore a black suit with a white shirt and blue bow tie. "Mildred!" he called toward the back of the store. "Mildred, please come and meet Peter Mertens!"

Martin introduced his Irish-born wife who was slender and somewhat taller then her husband. At forty she was very attractive, her blondish hair done in a pageboy manner, with straight bangs over her forehead. Her bright eyes were beautiful, her light make-up expertly applied. "What a woman!" thought Mertens. She wore a pinkish white suit, which was embroidered with silver threads. A pair of pink, high-heeled shoes covered her feet. The trio sat down on three red velvet chairs.

"*Monsieur* Mertens, we have to change you into a *Choucroute*, but please tell us first the latest of the allied invasion."

Yet again, Peter related his latest impressions from Normandy. The couple was absolutely delighted by the good news. 'When will they come to free Paris?" was the urgent question.

"Soon, in one or two months," Mertens guessed.

In the back of the store were two rooms, one an office, the other a sort of storage area where three suits and a light colored, lined trench coat hung inside an antique wardrobe. The shelves contained shirts, ties, socks, shoes and all the necessary paraphernalia for a traveling, German businessman, of Mertens size, including a dark blue hat. Mertens was impressed by the expert preparation and order. He left his bicycle and farmer outfit with the couple, after changing into a dramatically different person. Martin called a cab to his store. 'We'll stay in touch!" said Mertens, walking out from under the gold lettered "Gallerie Martin", then entering a most surprising vehicle. It consisted of a bicycle, which was coupled to an open box type crate, supported by a two-wheel axle. Inside the box was an upholstered bench for two passengers, but it was even more surprising, that the driver tuned out to be a young attractive woman. After Mertens had placed the suitcase inside the cab he asked: "Do you think you can take me to the *Hotel de Crillon*?" To his great surprise and amusement the girl managed the vehicle expertly. First she stood on the pedals until the vehicle overcame its inertia and moved slowly ahead over the pavement. Then she increased the speed and had no problem to keep up a good pace until they reached the hotel.

After tipping the woman well, he carried his suitcase, a brown floppy leather briefcase and trench coat over his arm, into the lobby of the hotel. At the reception desk he requested a single room with bath. The clerk asked for his German passport and after a short glance said. "*Monsieur* Mertens, I am sorry I can't give you a room until you have obtained permission from the German City Office for *Traffique* in the *rue de Berri*. It is a side street off the *Champs Elysees*." "May I leave my belongings with you?" asked Mertens. "*Certainement Monsieur!*" replied the clerk.

He walked the few long blocks uphill over the *Champs Elysees* from where, in the distance on top of the inclined avenue, he saw the outlines of the *Arc de Triomphe*. After a swift walk of about twenty minutes, he reached his destination and was obliged this time to speak German to the attending clerk; to whom Mertens showed all his pertinent documents and explained the reason for his stay in Paris. The German official was obviously impressed by his credentials and his explanation of receiving and safeguarding valuable art treasures for the *Vaterland*.

Returning to the hotel Mertens had no problem obtaining a key for his single room with bath from the clerk. In spite of the elegance of the lobby, his room on the third floor was fairly modest, although two windows provided a superb view over the *Place de la Concorde*. His first action was to take a bath and remove the dust and sweat from his lengthy bicycle trip. Now he felt more comfortable to wear his elegant German attire.

It was past noon when he decided to have lunch. The hotel restaurant was decorated with painted green ivy plants over white walls, and most of the dining tables had glass tops with chairs made of white enameled wrought-iron. He sat at a small round table with two·chairs and ordered a *Croquet de Madame*, plus soup from the waiter. The toasted sandwich arrived, covered with a cheese-like substance, that didn't taste too bad. He washed it down with a glass of plain soda water.

Now at last he had a chance to view his surroundings. Most of the guests were German army officers, together with a few French collaborators in their black Nazi *Milice* outfits. But he had to find a specific officer in SS-uniform. After the meal, he placed himself in a lobby corner from where he could observe the guests, including their wives or girlfriends.

Sitting in another upholstered chair immediately next to him, was a stocky, almost totally bald gentleman. He wore a dark suit with vest and a silk tie fastened by a prominent diamond clip. Mertens was enjoying the passing parade, when he asked his neighbor in German, if he was also staying in the hotel. The man answered in a flawless Cologne or Düssefdorf dialect: "Ja, I am living here already one week." Mertens introduced himself and explained the cover story about his art mission.

"My name is Karl Banzer," the man said. "I am an orchestra leader, hired to entertain the German troops in town."

Soon they were deep in conversation, the main topic being of course the invasion. The more they spoke the more Peter took a liking to this interesting stranger.

'What do you think will happen? Will the Allies come to Paris? Do we have to get out of here?" were some of Mertens' leading questions.

'"The enormous amounts of armament and materials, together with air-superiority of these Allied-Forces are a great threat to us. We can only hope upcoming German secret weapons can turn the war in our favor." Banzer said, looking Mertens straight in the eyes.

"Yes, I've heard of one of these secret weapons. I understand it consists of a rubber U-Boat, which will skirt the coast of England. After rounding the island a few times England will be completely erased." The man roared with a hearty laughter. The ice was broken.

Late in the afternoon Mertens retuned to René Martin's store and reported his successful settling into the hotel, he told also of the meeting with Karl Banzer. René replied: "Be careful, the place is riddled with *Gestapo*. By the way, I saved you a bottle of Napoleon Cognac; take it, it might come in handy."

Mertens walked back to the hotel where, shortly after returning to his room, there was a knock on his door. Two men asked to be allowed in. Both wore brown, shiny leather jackets.

'We would like to talk to you and see your credentials!"

"No problem, please come in!" replied Mertens.

The visitors entered and introduced themselves as Herr Wagner and Herr Schmidt. Mertens offered both drinks from the cognac bottle, which they greedily accepted, then he presented his documents to be inspected in minute detail. The military passport contained a stamp: 'KV,' which meant usable for war service, plus the markings 'Nzv', meaning, 'not to be used in accordance with a specific secret order,' followed by a lengthy number. Inserted in the document they found a separate certificate, stating Mertens to be a private investigator from a large Berlin firm. When he then explained the purpose of his mission, both men were impressed, telling him that if he needed any help, just to contact them in room number 234. As a final gesture Mertens handed them the entire bottle of cognac, explaining that he had a source for more where that came from.

After a few days of leisure and observation, either in the restaurant or in the lobby, Mertens met with Karl Banzer almost on a daily basis. More and more he was convinced, that Banzer loathed the Nazis.

Then, one day at breakfast, Mertens saw the SS-officer he was looking for. A tall, slender blond Aryan entered the hotel restaurant, wearing a gray uniform of the 'SS'. The insignia on his left collar indicated, he was a Standarten-Führer, the equivalent to an army Colonel. The right collar edge was marked with the white letters 'SD'. The Nazi officer was accompanied by a Schaarführer, the same rank as a Sergeant, who had the habit of carrying a stiff leather baton under his arm.

"Is it possible?" Mertens thought. "This Sergeant undoubtedly is the son of a bitch, who brutalized and kidnapped my landlord, his wife and daughter." Mertens became apprehensive. 'Will this bastard remember me or my name?" Getting hold of himself, he decided to ignore this danger and firmly play the part for which he had been chosen.

The two SS-Men sat down a couple of tables away, giving him ample opportunity to study their faces.

First he noticed the pronounced jaw of Kempe, making his pock-marked face look narrow and long. Then he observed the nose, matching the man's appearance through its long, fleshy shape. The steel-gray eyes too close together, resembled a caged primate, and his hair was a blond, kinky steel-wool mat. His uniformed companion was a typical stocky peasant, who had made the SS his life's career. His name was Horst Stolz.

While in St. Cloud with Bernard, Mertens had studied Kempe's background. The SS-Man was born 1915 in Dresden, Saxony, where he developed late. Studying law in Berlin, he joined the 'SS' in 1935. After his law graduation in 1939 Kempe advanced rapidly in the 'SS'. Eventually, Adolf Eichmann, Head of the *Gestapo's* Jewish Office, brought him into his group at Berlin's *Kurfürsten Straße*. Eichmann used Kempe for selected missions, including the secret U-Boat development, for which Kempe had to visit the Baltic Sea port of *Eckernförde*. Now here in Paris, he was in charge of rounding up French Jews, for shipment east to concentration camps. He also knew Kempe had a voracious craving for young girls.

Wagner, the *Gestapo* agent, entered the restaurant to keep an appointment with Kempe, and seeing Mertens, greeted him with a cordial Heil Hitler. Passing his table he told him, how much he had enjoyed the precious cognac. Mertens responded by asking Wagner: "Could I join you for breakfast?"

'Why not?" Provided the *Standarten Führer* has no objections!" answered Wagner. For a short moment Wagner talked to Kempe and then they waved: "Come and join us!"

Mertens, raised his right arm to greet Kempe and his cohort with a proper Heil Hitler, then shook hands and introduced himself. After the waiter had taken their orders, Kempe, speaking with a strong Saxonian accent, directed many questions at Mertens; who decided to reply in a well-Iearned Prussian-Berlin dialect.

"Where are you from?" asked Kempe.

"From Berlin-Wilmersdorf!"

"What brought you to Paris?"

"I'm an art expert, hired to receive certain artifacts from France and have them shipped to safety in Germany!"

"How interesting!" was everyone's reply.

Kempe asked: "Our Embassy has quite a few art objects stored. Are you by any chance involved in their activities?"

"No, not yet. My supply source is a French-German sympathizer who also owns a gallery in town."

'What's his name?" asked Wagner.

"Well, I don't want to embarrass my contact", said Mertens, "but he has an Irish wife, who hates the English swine with a passion. I am sure; we'll soon show these decadent Jew slaves, what Germany can do, once we bring all of our secret weapons into action.

"Bravo! Well stated!" said Kempe.

"I only know," continued Mertens, ''we Germans here in Paris have to stick together before these Frenchies get any wrong ideas!"

''That's for surel" Kempe agreed. The others echoed his opinion.

Sergeant Stolz quietly observed the conversation. He found this stranger who joined them acceptable, because he expressed the right belief in their present situation. However, he couldn't shake the idea that this Peter Mertens was familiar to him. Why and where he knew him, he could not recall. On the other hand he met many people in Berlin. For the moment, he decided to forget about his subconscious memory twitch, but he would tell Kempe about his concern, when he had the opportunity.

The next day Mertens was invited to join Kempe in his hotel room for a drink of cognac. When he arrived and entered the room, he noticed that not only was Kempe present, but also the two *Gestapo* agents plus sergeant Stolz.

"Please sit down! Hey Stolz, serve the gentleman a cognac!" said Kempe.

Mertens didn't like the look of this inquest-like assembly. Then Kempe started with his questioning. "Herr Mertens you said you lived in Berlin-Wilmersdorf. What was your address?" He was prepared, for back in Washington, it had been decided to pick a random Berlin family, with the name Mertens and use them as a reference. "I lived at Detmolder Straße 3 on the fourth floor and at the front of the apartment building. From our balcony I could look to the Kaiserplatz," was his reply.

"Where did you go to school?"

"I went to the 5th Public School on Koblenzer Straße and later attended the

Friedenauer Gymnasium at Maibach Platz."

Kempe, having lived in Berlin, was familiar with the two schools. Then he shot at Peter the next inquiry. "As an art dealer, where is your office and for whom do you work?"

"I have a partner in the Keith Straße, his name is Wilhelm Kuhn."

Now *Gestapo* agent Wagner entered the questioning game. "Herr Mertens, when were you the last time in Berlin?"

"I just came from there" was his answer.

"Then tell me about the latest bomb damage to the city."

Mertens was not prepared for this tricky question and feverishly raked his brain for an appropriate answer. He decided on a ploy, using great emotion, he forced an almost hysterical reply.

"These American gangsters and British swine have done a lot of damage to our capital city. When I left Berlin the smoke-filled destruction was still lingering over the city. I request that we don't hesitate any longer and use our secret weapons, immediately.

Our Führer, with his great wisdom will retaliate and guarantee victory for our great nation. Sieg Heil!"

Kempe stopped Mertens' tirade. "Herr Mertens we sympathize with you whole-heartedly. Please calm down; you must understand we have to be careful. Too many enemies and spies are lurking in this French city and we must make sure you are the person you say you are. Please have another cognac; we are finished with our interrogation."

Wagner was not at all happy with the sudden termination of questioning, for he just had been in Berlin and wanted to check Mertens on specific damage areas. He knew Kempe had the last word, but he decided, that sometime later he might question Mertens on his own.

From then on his new found companions breakfasted together almost every day. After a while Karl Banzer seemed to avoid Mertens' company, leading Mertens to think, that maybe his association with the *Gestapo*, had put a crimp in their relationship.

In early July, Mertens again visited René Martin's store, where a few surprises were waiting for him. He learned, that on June 27th the German garrison at

Cherbourg had surrendered, liberating the harbor, allowing an uninterrupted supply of US war materials to be unloaded. St.Lo was still in German hands, but the Brits appeared to be getting ready to capture Caen, starting with massive air attacks.

As Mertens reported his success in getting increasingly acquainted with Kempe, a big smile came over René's face. Becoming serious again he said: "The only fly in the ointment could be this damn Sergeant Stolz."

Then he continued: "I have quite a surprise for youl" René said. We have a luncheon appointment at my preferred cafe around the corner; I want you to meet a colleague of ours!"

The restaurant, which had an Alsatian owner, was located at the Place Vendome. The establishment was rustic with wall pictures of the Strasbourg cathedral, of storks sitting on church steeples in their nests, and of women wearing costumes with huge, white starched head-coverings. At a secluded corner table Mertens discovered none other than Karl Banzer, who with a sheepish grin extended his hand. They spoke French, Banzer's without accent.

"Karl has been working for the American OSS," explained René. "Right after you told me, I advised London about your meeting with Karl at the 'de Crillon', and to make a long story short, Steve McGuire, in agreement with the London office, decided that the two of you could work well together. So here is our plan of action." René explained the scheme in detail and everyone agreed to proceed accordingly.

During the next breakfast with Kempe, Mertens noticed a young French woman sitting at their table. She obviously had spent the night with the SS-officer. Kempe saw Mertens' hesitation to join him at his table, but he grinned triumphantly and imperiously dismissed his hooker, explaining that he wanted to talk to that man alone and in privacy.

After the prostitute had left, Kempe stared icily at Mertens. "I understand you have met my Schaarführer Stolz before?" he said.

Mertens was shocked by this approach, but with calm control he stared right back. "Not that I recall Herr Standarten F*ü*hrer; however, Berlin is a large city and maybe he has seen me in a restaurant or some other public place. Who knows?"

Kempe still staring, finally relaxed. "It just proves your story that you are from Berlin," he said. Peter decided to divert Kempe's attention from this dangerous subject by introducing a very tempting suggestion.

"Who knows how long we can stay in Paris"? he said.

"So before anything happens, we should have a monumental party. Do you know the bandleader Karl Banzer? Well, we have arranged for a secluded hall in a hotel on *rue de Castiglione* with Banzer providing a few musicians, and even girls from the *Moulin-Rouge*. I hope to be able to get hold of enough cognac and wine. Would you be interested in joining us?"

Kempe's face changed into a smirking leer and after a moment's silence answered.

"How many of our group would you say could participate?"

"I think we could accommodate ten defenders of the Reich," Mertens said.

"Wonderful!" exclaimed Kempe. 'We'll be there!"

CHAPTER 5

The party was set for July 19th. René Martin had taken care of renting the hall and a few rooms in the hotel on the *rue de Castiglione*. With the cooperation of French underground members the best available food, wine, champagne and cognac was brought to the hotel. Chairs and small tables formed a horseshoe pattern around the parquet dance floor; while at the far end there was a raised platform for the band and entertainers. The guests were invited to arrive at about ten p.m.

One hour prior to the party two-army trucks arrived at the hotel, and about twenty Waffen-SS-soldiers jumped from the vehicles. Some positioned themselves in front of the hotel; while a group led by Kempe's Sergeant entered the building, checking every floor, room and the dance hall. SS-Standarten-Führer Heiner Kempe wanted to be sure the invitation was safe. Further, he decided to have the soldiers stand guard, while the party took place, also to have some of the Gestapo members check the food and drinks.

René Martin arranged for the arrival of fifteen ladies. Most of them were patriotic prostitutes or line girls from local nightclubs. They had been given strict instructions for the upcoming evening program. The band consisted of twenty musicians, who were a mixture of nationalities, mostly from Italy, hired by Banzer. Normally, they entertain the German troops in and around Paris. They played all of the popular German dance music by Michael Yari, Peter Kreuder, Peter Igelhoff, Paul Linke and many others. Karl Banzer served not only as band leader, but also as Master of Ceremonies. He told a few jokes and introduced the women. One of the sentimental tunes, he played, included the lyrics: "Es geht alles vorüber, es geht alles vorbei!" ("Everything will go away, everything will eventually pass!") Waiters served the guests drinks and food.

Mertens spent most of his time with Kempe. He could see his guest enjoyed himself enormously. A young Frenchman and member of the band presented a few saucy chansons, after which the guests had a marvelous time dancing with the girls. Sometime after midnight all the girls disappeared. The MC announced a special feature attraction, and the band burst into a 'Can-Can'. Suddenly all fifteen girls appeared completely nude on the stage. The guests were flabbergasted, and then a great cheer was heard. Instantly, the lights dimmed and the girls ran from the stage into the audience, jumping on the laps of pre-selected partners. Most of the men, soon disappeared to the available hotel rooms, carrying the women in their arms.

Mertens' thoughts went back to Chicago, when he and two fellows from his senior high school class, went on an evening's adventure on South State Street, immediately next to the downtown loop. It was at the Gaiety Theater, where they eagerly watched their first strip tease performers. In between various girly numbers, they enjoyed a few comedians with their bawdy acts. The entertainment always ended with a strip-tease star performer, who revealed her qualifications only in the very last moment.

Quickly, the entertainment hall quieted and only Mertens, Karl Banzer with his band and a few waiters were left. Karl joined Mertens at his table and both agreed the party had been a success.

Suddenly a scream could be heard from the hallway leading to the rooms. A naked woman holding a bed pillow in front of her ran toward Mertens table. 'What happened?" shouted Banzer.

"This SS-man monster tried to rape me. He became brutal and hit me in the face and he beat me with his belt all over my poor body. I want to get out of here. Please help me."

The woman shrieked and begged in tears, interrupted only by her sudden sob attacks. Both Banzer and Mertens comforted her and tried to calm her down. From her description it became clear that it was Kempe, who behaved like the brutal beast, he was. Mertens fed her some cognac and handed her a couple of hundred Frank notes. Karl took her to the dressing room and delegated one of his musicians to escort the devastated lady home.

The next day, July 20th, Mertens slept late at the *'de Crillon'*. It was just after midday, when someone urgently knocked on his door. Bleary-eyed, Kempe stood in the doorway shouting:

'"They tried to assassinate our Führer with a bomb! A bunch of blue-blooded swine have plotted to take over the government; however I got word from Berlin that Hitler is alive, and the traitor bastards have been rounded up! I just wanted you to know, before I'll have to rush to headquarters! We're going to be very busy!" Before Mertens could say a word, Kempe was gone.

Amazed, Mertens closed the door and dressed. When he arrived in the lobby, the guests were gathered around a radio, listening to the latest reports in French. A Colonel von Stauffenberg had planted a bomb at Hitler's headquarters in East Prussia, and a number of staff members had been killed. Hitler, shielded by a heavy oak table, was only shocked and his arm-twisted by the blast. Hermann Göring gave a report over the radio, assuring the German people, that everything was under control, and the perpetrators captured. A search was on for a Karl Gördeler, who was slated by the plotters to become the new chancellor, after a successful uprising.

Mertens decided to hurry to René's store to see if he had more information.

"That son of a bitch got away again!" René shouted angrily.

After they reviewed the situation, they decided to continue with the original plan, regarding Kempe and his Berlin connections.

Back at the hotel, Mertens discovered a few of the German officers were missing; the rumor was, that the Gestapo had arrested them. The City Commandant General Karl Heinrich von Stülpnagel had also disappeared from Paris, after he had given a party at the hotel Raphael. Mertens knew at this point that he had to sit back and let the dust settle, before anything further could be done.

A couple of weeks went by, before Kempe and his *Gestapo* cronies appeared again at the breakfast table. When he arrived and sat down, he was exhausted, but freely volunteered the latest information.

On July 8th, St.Lo had been captured, after the town had been bombed to smithereens. Mertens knew now the stage was set for the Allies to break out of Normandy, in the direction of Paris. The new German commandant in Paris was now General Dietrich von Choltitz.

Kempe had it from reliable sources that German secret-weapons would be used at any moment, resulting in the war, definitely turning to Germany's advantage. Already the 'V-1' rocket was being deployed regularly against Britain and other much more significant weapons would follow. Kempe even mentioned a bomb, which could wipe out an entire city like London. Then he started raving against the Jews, who in his opinion were the instigators of war.

"This afternoon I have to supervise a deportation of French Jews from France to their deserved destination in Eastern Europe," Kempe said.

"As Dr. Göbbels once said: 'Should we ever be forced to leave the stage of world events, then we shall slam the door in such fashion, that nobody at all could be called a Victor!'"

Mertens saw it as his duty to witness Kempe's activities, so he asked: "Do you have any objections if I would join you on your aftemoon's mission?" This request startled Kempe.

"If you want to see a bunch of human vermin, please be my guest."

He advised Mertens to report to him at 2 p.m. at the RSHA (*Reich's* Security Main Administration, Central Office), Branch for Jewish Questions at *31 avenue Foch.*

As agreed, Mertens met Kempe at 2 p.m., where a contingent of French Milice personnel was assembled with their trucks and cars, under the command of *Commissaire de Police Francois.* Mertens joined Kempe alone in his Mercedes. The two drove off heading for *rue de Flandre* in the direction of LeBourget. Right behind followed Kempe's Sergeant on a motorcycle, with sidecar.

Kempe became jovial and very talkative. "Even if the war should be lost, the Führer has demanded a select few of our great party organization will carry on, spreading the word and continuing actions against the Jews.

"We have already made excellent contacts in South America," he continued. One of our secret weapons is a number of special submarines that are capable of staying submerged for many days. They are so fast that no Allied vessel can catch them, even while the sub is traveling submerged. Don't ask me how they work, it's top secret. Believe me, these U-Boats do exist, I saw one with my own eyes."

Now Peter understood why he was ordered to stay close to Kempe. The possibility of these Nazi thugs escaping, while continuing their atrocities all over the world, was absolutely unthinkable. This plan had to be crushed, before it even got started.

After a short drive, the car reached the small district of Drancy; where they stopped at a large area, enclosed by a high fence and barbed wire. Watchtowers with machine-guns overlooked the enclosure, but surprisingly, the whole encampment was skirted by apartment buildings. Kempe parked the car in front of the gate and said, "We'll have to wait for the rest of our men to arrive!"

Mertens could see the inmates of the compound, most of whom were well-dressed, consisting of families with children, all wearing a yellow star on their left front, even the older children. In Mertens' opinion, to consider these harmless people dangerous or violent was completely insane.

At one side of the compound, he noticed a railroad siding with maybe twenty freight cars, standing ominously on the tracks. Soon the Milice trucks arrived with about thirty French collaborators, armed with machine pistols. After opening the freight-car doors, they created a soldier's corridor, connecting the camp with the rail siding. Another ten men entered the camp and herded the subdued prisoners through the corridor, toward the freight cars. Many of the adults carried suitcases, but were instructed to leave them behind. Using the butt-ends of their weapons the *Milice* pushed, and in some cases brutally struck the defenseless victims, as they struggled forward. All removed luggage was neatly stored on to one side.

Especially brutal was Kempe's Sergeant, who used his baton like a whip, flailing from right to left striking men, women and children. Mertens' blood was boiling and he mentally made a notation of the Sergeant's motorcycle license number 'SS-A 231'.

To Mertens' horror, he noticed that each railroad car was being jam-packed, providing no space for the people to sit or spread. The children started to complain, and the young babies were crying, clinging to their mothers in utter panic. One young man started to resist the brutal treatment by pushing the Milice man away. Furious, Kempe ran to the Jewish lad, and shot him in the neck. Mertens' was devastated. He wanted to kill Kempe on the spot. It took an enormous amount of self-control to stay cool, not revealing his turmoil and feelings of anguish.

On the way back to Paris, Kempe appeared completely cool, he acted as if nothing special had taken place.

"Now you have seen how we deal with Germany's enemies," he said.

"I don't understand how these harmless-looking civilians could be a threat to Germany?" Mertens replied.

"Obviously, you haven't been educated and schooled, regarding these Jews. I have attended various Party-sponsored classes and special courses where they taught us about the evil of International Jewry. Believe me, my comrades and I know what we are doing, and one day the world will be grateful to us! Maybe it was a mistake to have taken you along today."

"Just the opposite, I am thankful for the experience. I have learned a very important lesson."

Mertens swore to himself that he would never forget this evil man. He vowed that he would revenge the death of that helpless, innocent boy.

CHAPTER 6

Toward the end of July, the situation in Paris became precarious.

Frequently, the electricity was cut-off and Mertens noticed the German army personnel starting to pack their belongings. General Patton and his Third Army had taken Avranches on July 30th, and now he was moving up the Loire-Valley toward Orleans. Clearly it would be only a matter of days, before Paris would be liberated.

It was timely for Mertens and Karl Banzer to visit their information source, René Martin, who was always eagerly expecting them.

"I received an urgent message from Bernard in St.Cloud. He wants to see us immediately," said René.

Leaving Mildred to manage the store, both men took the Metro-Subway to the end of the line in Boulogne, from where they walked across the Seine Bridge to reach St.Cloud.

They had telephoned ahead so Bernard was waiting for them. He took the two men into his office and asked them to be seated. Standing, Bernard folded his hands behind his back. Now he explained the latest instructions from London.

"As you can imagine, Mertens' information about SS-Standarten-Führer Heiner Kempe's remark, regarding secret U-Boats, and its planned usage for an escape to South-America, stirred serious reactions in London. Can you imagine the Nazis creating a Third Reich in Argentina or Paraguay, from where they would spread their evil conspiracies to other countries?"

"Yes I could imagine them capable of all kind of evil deeds," said Mertens. "For instance, a raid on the Panama Canal by blowing up the locks." Bernard nodded but then continued with his latest information.

"Through Polish agents we have learned, that the Nazis have tested a newly developed submarine at Hela near Danzig. But at this time we don't know where these vessels are located. However, somehow we must find and destroy them! We only heard they might no longer be at Hela!"

"That's exactly why I have to stick close to Kempe," said Mertens."

Bernard raised both hands to calm Mertens down, and then he interjected: "London suggests that Mertens and Karl intensify their relationship with Kempe who hopefully can lead them to the location of the U-Boats.

Ideally, you two, together with Kempe, should make your escape from Paris east, toward the German lines. This scheme would help to strengthen your relationship with Kempe. With that in mind, we have developed a plan which you should memorize and follow assiduously."

At this point, Bernard went into a lengthy and detailed plan, providing timing, location and method for the stratagem. Karl and Mertens had to repeat all of the instructions exactly. It was decided that Mertens, after having spoken to Kempe, would notify René about the date and timing for each of the individual phases of the plan. An obvious feeling of excitement got hold of these men and they soon broke off their meeting.

As usual at breakfast in the hotel, Mertens and Karl joined Kempe with his two *Gestapo* thugs, Wagner and Schmidt. The conversation revolved around the evacuation of Paris. Kempe said that Hitler had given specific instructions to General von Choltitz, demanding Paris be destroyed and burned.

"It is now up to the SS-Corps and the *Gestapo* to make sure these orders are carried out. Consequently, before my leaving Paris we will be required to enforce last-minute actions."

Mertens and Karl promised they would be available for any assistance. 'We Germans have to stick together!" Karl said.

Kempe always parked his Mercedes in front of the 'de Crillon' at the Place de la Concorde. Accordingly, on August 10th, the first phase of the London plan was put into action. During the breakfast hour, two French agents and expert car mechanics, broke into Kempe's parked Mercedes and stole it. After the discovery of his car theft, Kempe just went berserk. He smashed the coffee table in front of him and threw his chair against the wall. He swore to get even with all French vermin by killing them, knowing precisely who the enemies of Germany were.

His new problem was to find transportation. All of his colleagues had their vehicles organized. It was essential to find a replacement for his Mercedes, or in the worst case, hitch a ride with someone else, because transportation was becoming critical. Riding in Horst Stolz' sidecar was below his considerations. Anyhow he had other plans for himself.

A few evenings later, phase two of the London plan went into action.

Gestapo agents Wagner and Schmidt were packing their belongings to get ready for their departure from Paris the next morning, when two uniformed SS-men knocked at their door and entered.

"We have a car waiting downstairs. We are under orders to help you get out of here. SS-Sturmbannführer Herbert Hagen, prior to leaving Paris, wants to see you at the Hotel Scribe together with all of his men.

We noticed, that the so-called French underground has started to snipe at our people. We already have had a few casualties. These treacherous cowards must be dealt with."

Both Wagner and Schmidt had never seen these men before and demanded to see their credentials."Of course." The two callers stepped toward the *Gestapo* agents, and instead of producing their documents, expertly thrust their trench-knives into the chests of both Nazis. They uttered muffled cries but were dead before they hit the hotel room floor; their pulsating blood drenching the elegant carpet.

On Thursday August 17th, 1944 and during breakfast with Mertens and Karl, Kempe the SS-Standarten-Führer, made the following announcement:

"The German army is pulling out today! Most of them will leave by train from Gare de l'Est and Gare du Nord! I'm supposed to join Commandant Brünner at the Paris Bobigny railroad station, where fifty-one more Jews are to be sent east, under our custody. I gave my driver Horst Stolz instructions to leave Paris on his motorcycle and wait for me in Strasbourg at the hotel Maison Rouge."

Suddenly he was interrupted; the hotel manager appeared at their table. "Pardon Monsieur *Standarten Führer*, but I have terrible news! In room number 234, two German gentlemen obviously have been murdered."

Kempe knew exactly who these two men were, for he already wondered why they hadn't shown up for breakfast. Kempe completely lost his composure when he heard the news, ranting and raving and swearing mighty revenge. He was utterly furious and together with the manager went upstairs, followed by Mertens and Karl. Kempe in his almost insane emotions identified the murdered men and with a cold determination, completed his inspection of the gruesome scene.

He then went to the telephone to call the French police; but in spite of many attempts, he couldn't get a connection.

Finally, in the corridor, he got hold of two German army officers, who were on the way to leave town. Instead, he brought them to room 234.

"I order you to immediately round up twenty French hostages and have them shot in retaliation for the murder of these two innocent German citizensl" He screamed his command at them. He demanded their names and unit designations, promising to check if his orders had been carried out.

Both officers assured him they would immediately report this demand to their superiors, for prompt follow-up. Kempe requested from the hotel manager to have the bodies of the two *Gestapo* men removed to the city morgue. The Frenchman promised to comply as soon as possible.

However, with great apology, he informed Kempe that the Parisian police had gone on strike.

Mertens suggested that Kempe should change from his SS-uniform into civilian attire. Since both had about the same shape and figure, Peter offered him one of his suits.

"Today you can't walk through the streets of Paris in a German uniform, without getting into trouble," Mertens advised him.

Kempe rejected the idea of losing his uniform and status, plus the power it represented. He refused any further discussion.

On August 23^{m}, the latest news report stated that the allied forces were at both sides of the Seine River, namely near St.Cloud in the west and St.Maur in the east. Continuous shooting could be heard from the streets of Paris and it even had spread to the suburbs, encouraging the now spontaneous French resistance to become more organized, and take to the streets. With flaunted arrogance, Kempe still appeared in uniform the next day at breakfast. "I am not leaving this place until I am assured my comrades have been avenged," he said. Karl and Mertens replied: "We agree, but let's not miss our chance to get out of this crazy city!"

During the evening hours Mertens and Karl went to René's store. Here they learned that Eisenhower had decided that French General Jacques Leclerc's 2nd Armored Division, together with the US 4th Infantry Division would enter and liberate Paris the following day. René confirmed, that phase three of their plan would go into effect, the next day between 11 a.m. and noon.

The city was active all night, with the French resistance becoming bolder and more aggressive, causing German army units to leave town in hurried disarray. Wagner's and Schmidt's bodies were still in room 234 and Kempe had lost contact with his unit, without knowing if his orders of revenge had been carried out.

Late in the morning, Peter and Karl met Kempe again in the restaurant, but no hotel personnel had turned up for work. The manager was missing and the three felt quite conspicuously vulnerable. Kempe wanted to make his way to the Bobigny Station, but time had run out for him. While they were contemplating their options, they noticed American Sherman tanks driving over the Place de la Concorde. The Allies had arrived. Thousands of Parisians were all over the area, climbing on top of the tanks, screaming, yelling, and throwing flowers, waving American and French flags, fraternizing with the liberators of their beloved Paris.

Mertens and Karl decided to escape from the hotel immediately, suggesting to Kempe that he seek refuge at their art-dealers place on rue du Faubourg St.Honoré. Now phase three of the plan went into action.

Hurriedly, they left the hotel through a side door, with nobody bothering them. Kempe wore Mertens' trench coat over his uniform and hid his cap under the coat. They reached rue de Royal, where a Sherman tank stood in front of a restaurant. The crew, including its commander, had left, celebrating the liberation with surrounding citizens.

Mertens alerted quickly the other two and running over to the tank jumped into the right front opening. Karl entered the left opening and the startled Kempe climbed into the top turret. Before anyone could get to them they closed the tank covers, sealing off all interference.

Mertens knew where the light switch was located to illuminate the darkness and he readily found the starter-switch to his left. Twisting the switch, the still-warm aircraft gasoline engine started immediately. Looking through the periscope, Mertens saw his way clear. With his left foot he pushed the clutch pedal. Then he engaged the gear lever on his right, shifting it into first. Gunning the engine with his right foot, he pushed both steering levers slowly forward. Disengaging the clutch, the tank started to move. Shifting from gear to gear, the vehicle moved steadily forward. Beyond the *Madeleine* church they reached the *rue de Tronchet.* Then heading north over the rue Amsterdam they passed a short stretch of *avenue de Clichy*, until they drove into *avenue de St.Ouen.*

Straddling the steering rods with his legs, Mertens carefully directed the vehicle, by pulling the levers back, either to the left or right. This action slowed either side of the caterpillar tracks providing the desired steering action. In fifth gear, heading north they reached a speed of 25 miles per hour. Mertens drove east on *boulevard Ney*, until he arrived at *rue du Flandre*, then he continued north in the direction of Cambrai. Changing course again, they entered the desired highway to Soissons. "Where are the Germans lines?" He wondered.

Kempe had found a pair of earphones, as had Mertens and Karl, so they were now able to converse with each other without being hampered by the engine noise. To make the tank look less aggressive, it was decided to turn the turret, leaving the gun barrel pointing to the rear. Kempe located the hydraulic switch right next to him, making this maneuver possible. Mertens checked the fuel tanks on his left instrument panel, and found both still three-quarter full.

Once out into the open country, Mertens stopped where nobody was to be seen. Now they opened all hatches. Kempe enthusiastically congratulated Mertens on his audacity. "Where in the hell did you learn how to drive this tank?" He asked.

"A good friend of mine serves in the *Panzer-Corps* and told me how these things function. They are quite easy to handle, since it is actually a duplication of any automobile; the only difference being the arrangement and usage of the two steering levers."

"That was a hell of a stunt you accomplished. I owe you a lot for this ingenious escape," said Kempe. "A guy like you belongs to our brotherhood, the SS-Corps, and I'll talk to my superiors when I get back to Berlin. We need you!"

Mertens was shocked by this turn of events and the suggestion to become part of these murderous gangsters, seemed unthinkable. He had to ponder that one over carefully. Changing the subject, he suggested they find a straight tree branch to use as a flagpole. After selecting and cutting a suitable length, Mertens removed his white shirt and knotted the sleeves to the limb. Kempe stuck the sign of surrender above the turret, and with open hatches they continued their drive east toward Reims. Approaching the city, they noticed a rear guard of German soldiers on the outskirts. Kempe, now in his full uniform waved his arms and shouted: "We are Germans, don't shoot!"

After reaching the German lines, the trio abandoned the tank and asked to speak to the area commanding officer, who became very helpful after seeing the rank insignia on Kempe's uniform.

A Captain Hermann Bräuer established their identities, after which Kempe requested transportation to the next German town. But Mertens immediately suggested traveling onward to Strasbourg, as he had important, high level art dealings to settle there. The German Captain replied, that he had a messenger driver, who would leave for Nancy; but from there they would have to seek new transportation.

The driver, a corporal, drove the trio in an Opel-Kadett. The car roof was camouflaged with green tree branches. The weather was perfect, sunny and clear, which meant that Banzer and Mertens had to keep their heads out of their side windows, constantly watching the sky. Twice they had to stop and seek shelter in ditches, due to strafing Thunderbolt activity.

Finally, they reached Nancy, where Kempe hurried off to see the city Commandant, leaving Mertens and Karl in a cafe, where they eventually found something to eat.

After a while, Kempe returned, saying he had been able to arrange for a special car and driver to take them immediately to Strasbourg. Mertens and Karl were elated. Everything had turned out the way they had planned.

CHAPTER 7

Late in the evening of August 25th, Kempe, Banzer , and Mertens, driven by their chauffeur, arrived in Strasbourg, where they immediately went to the hotel *'Rotes Haus'* (Red House), located on the west side of a large square in the center of town. The hotel was constructed of red sandstone, explaining its name. After 1940, and the German annexation of Alsace-Lorraine, the French word 'Maison Rouge' was simply translated into German.

As in Paris, Mertens and Karl Banzer had to register the following day with the German Traffic Office, in order to obtain permission to stay in the hotel. They were provided with single accommodation permits.

Mertens' room was on the third floor; a large inward-swinging window permitted a view over the court yard and below where the glass roofed kitchen facility. The room had one single bed, a desk with chair, a wardrobe, and a wash basin with mirror. On the wall between the door and the washbasin hung a telephone.

The next morning after a good night's sleep, Mertens went downstairs to become familiar with his new surroundings. He found the lobby could be reached from the main entrance and the center stairway. Large floor to ceiling windows overlooked the *'Rote Platz'*, formerly called *'Place de Kleber'*. A few ceiling-height potted palms stood near the bright picture window area. The lobby with its oriental carpet, on tile flooring was decorated in a typical French salon style. Large mirrors covered the walls, including a sizable tapestry, picturing a Baroque shepherd scene. The ceiling lighting fixtures consisted of two prominent crystal chandeliers. Adequate leather seating, with coffee tables contributed to the comfort of the guests. Directly adjacent to the lobby was the dining room, separated by heavy brocade draperies.

The most conspicuous guests were the French-speaking collaborators, with their wives and girl friends; now fleeing the wrath of their countrymen. Some of them still wore black uniforms with shiny jack boots. They all traveled in their ominous black 'Citroen' sedans, several of which, were parked at the front entrance, still covered by green tree branches.

Mertens walked into the lobby, found a comfortable wing chair and watched everyone who entered and left the hotel. After a while, a tall slender fellow sat down next to him. Pockmarks covered his sallow face. He had big hands, and his fee when he walked, were flat, spread outward. Mertens didn't like the man's looks at all.

Suddenly the stranger turned to him and said, "My name is Krause! You must have arrived just recently?"

Mertens answered: "Yes, we arrived last night from Paris."

At that moment Heiner Kempe walked into the lobby, heading straight for Mertens and Krause.

"Mertens I want you to meet Herr Krause, who is one of our men. He and another agent, Ernst Hochstetter, are assigned to this hotel, just like our murdered friends in the *'de Crillon'*; Wagner and Schmidt. That reminds me, I must check if those officers back in Paris followed my orders by executing the twenty hostages."

Looking at Krause, Kempe explained to him that Mertens and Karl Banzer were the ones, who helped him escape from Paris in the Sherman tank. Krause was greatly impressed and congratulated Mertens on his nervy caper.

Kempe went back to his former thoughts and decided to make sure his vengeance was carried out.

"Krause, I want your men to pick twenty French prisoners from the 'KZ' (concentration camp) Schirmeck, and have them executed for the murder of our comrades in Paris, now; before the Amis (German demeaning slang for Americans) get to Strasbourg. I'll instruct Schaarführer Horst Stolz to work with you.

Turning to Mertens, he continued: "Today I'm leaving for Berlin. I've received orders from my chief to be there within two days. Look me up when you get there. You'll find me at the office of Adolf Eichmann in the Kurfürsten Strasse."

Mertens was still apprehensive about Stolz, hoping he would not remember their former encounter in Berlin. He had to find a way to make sure, to eliminate this dangerous man. Then, his thoughts returned to Kempe and he replied. "First I have to take care of some art transfers here in Strasbourg, but afterwards, I'm planning to come to Berlin to finish my work assignment. I will certainly look you up, and we'll get together again. Have a good trip; I'll see you soon!" After shaking hands the two parted company.

Before leaving the hotel, Mertens grabbed the house phone, inside one of the two wooden phone booths, and called Karl Banzer's room.

"Hello Karl, I am on my way to see Eduard Colline; you know, our man in Strasbourg, you got his address from René in Paris. Meet me there in half an hour!" Karl acknowledged the request and promised to be there.

Mertens walked north toward the Kellermann Kai, a street protected by a black iron railing, along a side arm of the river Ill. He crossed over the bridge, eventually reaching his destination in the Bouxviller Straße. It was a part of town, consisting of solid apartment dwellings, built around the turn of the century, undoubtedly for the upper middle class. The pleasant walk from the hotel had taken about fifteen minutes in warm weather under a blue sky.

Bouxviller straße number 12 consisted of a four-story apartment building, with a light gray facade, wide windows and sepia brown rolling shutters.

Pushing one of the many bell buttons, outside of the entrance door, Mertens announced his arrival to Mr. Eduard and Mme. Lucie Colline. After climbing three floors, he reached the open apartment door. Since it was Sunday morning *Monsieur* Colline was home. During the week he spent most of his time in the suburb Schiltigheim, where he and a partner owned a shirt factory. "My name is Peter Mertens I am looking for 'Blondie'!"

"Jean Harlow isn't here, but please come in, I am expecting you!" replied Colline, a small stocky man of about fifty.

Mertens noted his French had a typical Alsatian accent, so he decided to continue the conversation in German. The couple showed extraordinary hospitality, starting with a glass of Pernod mixed with some water, which was followed by a bottle of champagne.

The couple's first questions were of the latest news, regarding American military actions. Mertens recounted as much good news as he could.

"The present American front line was halted at the western edge of the Vosges Mountains near Luneville," he concluded. He also told his hosts that Karl Banzer would join them in a few minutes, explaining:

"We have a lot of work ahead of us and we need your help. You'll have to introduce me to your friend Dr. Rasche, for I know it was you who reported his existence to our 'OSS' group." At that moment the doorbell rang and Karl Banzer arrived.

After introducing him, Mertens explained that Karl was to establish a radio trasmission station in the Vosges Mountains, which had to be able to communicate with the allied forces in Paris and the American Consulate in Zürich.

"The radio transmitter and receiver station will consist of a bread-box shaped unit with earphones, to be hidden on a small wagon, loaded with lumber and drawn by a couple of horses!" It is a similar ploy like the one Bernard used in St.Cloud. One of the larger diameter tree trunks will be hollow, providing space for the insertion of the transmitter. A carpenter has to be found, who would fabricate this bread-box size cavity; including a lid for the same tree trunk, making it hopefully impossible to be discovered even in the case of search!"

Eduard Colline became excited when he heard about this plan. "What a clever idea; I can't wait to get started with this work!"

Karl continued: "The problem is to obtain an efficient antenna for the radio; however, I believe I have found the solution. At three predetermined spots, maybe twenty kilometers apart, in the forest, we'll find a tall elm tree covered by ivy vines. We'll fabricate an antenna from copper wire with a dark brown insulation, to which we'll attach ivy leaves. This wire will be strung to the highest point of the tree, where at the top it will give maximum range. Every time we need to transmit a message, our wagon will be brought to the foot of the tree and the pre-installed antenna will be connected. Then, by moving between the three locations we'll keep the Germans guessing."

"I have the right carpenter for this job!" Eduard Colline replied enthusiastically. "He is located at a small village called Still, right on the easterly foothills of the Vosges mountains, where by the way, our secret printing press is also positioned!"

Turning to Colline, Mertens said: "As soon as we are finished with our meeting here, would it be possible to see Dr. Rasche?"

"Of course, my time is your time!" Eduard answered.

Mertens continued: "We have one nastier problem remaining to deal with. In Paris our people exterminated two *Gestapo* agents and now, this SS bastard Kempe has ordered, to take hold of twenty Frenchmen at the Schirmeck concentration camp, and have them executed in retaliation. He has commanded this to be carried out by his Aide Horst Stolz and two *Gestapo* agents, named Paul Krause and Ernst Hochstetter. Both occupy a room at the Rotes Haus. Somehow we have to stop this cold-blooded murder!"

Eduard said he would contact his underground organization to have them find a solution. "In fact, hidden with Dr. Rasche are two French officers, who would be just the right persons to handle this affair!" he explained.

After the meeting's conclusion, it was agreed that Karl Banzer arrange immediately for the fabrication of the antenna. Colline called an electrician friend in town telling him to meet Banzer in half an hour and that Lucie Colline would bring him over to introduce him. Eduard said the radio transmitter would arrive with an agent coming from the American lines, who would deliver the unit to the secret printing place in the village of Still.

Leaving the apartment, Eduard Colline and Mertens walked a few blocks south, past the Rotes Haus, arriving at the Ketten Gasse number 2, formerly rue du Chaim. The area consisted of old buildings with quaint facades, providing a romantic charm. A round arch-shaped gateway led into a small courtyard. A back door guided them into the building's right wing. An old wooden staircase, its steps worn over the years, brought them to an apartment. Colline rang the bell on the second floor; Suzanne Abel, the apartment owner, opened the door, recognized Colline and invited the pair to enter. After reassuring herself that nobody had followed them she closed the door.

"Eduard, who is this handsome young stranger?" Suzanne asked.

"May I introduce Peter Mertens, who would like to meet the Professor. He has come a long way for this purpose and I believe the Professor is expecting him."

At this point Mme. Abel knocked on one of the corridor doors: *"Herr Professor Sie haben Besuch*" (You have visitors)!"

The door opened. A tall man stood in a half-lit room. In a deep booming voice he said. 'Who want's to visit me? Please come in to my modest quarters!"

"My name is Peter Mertens, Herr Professor. I have come to talk to you about hydrogen peroxide and plans for your future!"

Suddenly, the professor became overcome and speechless. Mme. Abel silently left the room and closed the door.

"May we sit, Herr Professor?" Colline suggested.

"Please make yourselves comfortable in the two arm-chairs!"

Mertens noted the room was L-shaped. The open curtain in the corner revealed a neatly made up bed.

The professor seated himself on his couch-bed. After a pause he said to Colline, "Eduard you really have kept your promise, I only hope our efforts will not be too late. The Nazis must be stopped, before using their most deadly *"Wunderwaffen"* (Miracle-Weapons).

"Professor, the outcome of the war is no longer in question!" Mertens said. 'We are however concerned that a group of Nazi war criminals may be able to escape to South America via these radically-designed, new U-Boats. From there they could be spreading their perverted doctrine all over the world. That's why we must stop them, by destroying these dangerous submarines. I'll need all the technical information you have, so we can take effective counter-measures!"

At that moment Mme. Abel knocked on the door and brought a tray with coffee cups and a large pot of imitation coffee, together with a number of homemade cookies into the room. After she left, Mertens told the professor about his mission.

"Herr Professor we must change your identity! By the way, where is your daughter Erika?"

"Oh she went for a walk, but she should return any minute. It isn't easy being cooped up in these quarters day after day!"

"I understand", responded Mertens. "As I started to say, we must change your paper identity into Professor Dr. Ernst Berger, art expert and art curator. Your daughter would maintain her first name and should be registered as a nurse. I believe she studied to be a medical aide in Berlin?"

"That's correct", replied the Professor.

Mertens continued: "I believe Eduard should explain our urgent plan of action!"

At that moment, the door opened and Erika stepped in to the room. In spite of the poor lighting, Mertens was stunned by her beauty. Her wavy light blond hair fell casually over her shoulders. Her aristocratic face, marvelous blue eyes, and lush sensuous figure left him breathless. He was struck like lightning and remembered seeing her before in Berlin. Eduard introduced him while Mme. Abel added another cup to the coffee service.

Eduard explained that on the next day, he, the Professor, Erika and Mertens should take a bicycle excursion trip to the village of Still, where they would also receive new identities.

At this point Mertens interrupted. "What we want is to get the Professor into Switzerland, as soon as possible, provided he'll agree. However, I still have to make more arrangements and prepare a detailed plan. Tomorrow, on the way to Still, the Professor can explain the technical aspects of the H_2O_2 propulsion system to me. At 7 a.m. we'll meet at Eduard's apartment and begin our trip from there. Eduard, I believe it's time you and I meet with the two French officers, provided they are in the apartment."

Mme. Abel led Mertens and Eduard to the end of the hallway, where she knocked on one of the doors. *"Entrée!"* a voice called out. They entered a large room, containing two single beds on opposite walls; a table and four chairs stood in the middle. Two bare-footed young men, dressed in dark pants and white shirts, rose from their beds. The room reeked of the Gauloise cigarettes still dangling from their lips.

Eduard of course knew them, and introduced Mertens as 'one of our comrades from America'. They presented themselves as Lieutenant Pierre Schneider and Captain Didier de Roos. Both were of medium height. Schneider had blonde hair and obviously, due to his lingo was born Alsatian. The dark haired de Roos came from Metz, north of Lorraine. First they spoke French, but then Mertens switched to German, to discover if they spoke it well; both spoke it fluently; although Schneider's accent held a slight inflection of the Alsatian dialect, somewhat like Swiss-German.

After Mme. Abel had left the room, they sat around the table to get further acquainted. As usual, Mertens had to tell all he knew of the Allied war progress and the latest news from Paris. Finally, Eduard reminded them of their visit's purpose and related the murderous 'SS' intention to execute twenty Frenchmen from Schirmeck. After many questions and suggestions, the two officers agreed on de Roos' plan of action. He then explained:

"First, we must keep an eye on the *Gestapo* man Paul Krause and learn what he's up to. We have just the right person who can keep tabs on him; he is renting a room next door. His name is Werner Garewski, a sixteen-year-old youngster from Germany, who hates the Nazis. Since he refuses to be drafted, he ran away from his hometown Lahr, and is now waiting for the Americans to arrive. We have charged him to mingle and keep an eye on the French collaborators in the

Rotes Haus. I believe he knows Krause and has made good friends with some of the French girls in the hotel. We can trust him. All he has to do is to inform us when Krause is leaving the hotel. Our men will be watching from the Place Kleber and should he or any of his SS-men leave for Schirmeck, we'll be prepared to follow and take care of them. As a matter of fact Werner is here; he left us just a few minutes ago.

We'll ask him to join us, since I want you all to meet him."

At first Mertens didn't like the idea of involving a German boy in this dangerous operation. He was concerned, that if the youth was trying to avoid the German draft, his identity papers would possibly be inadequate to protect him from the *Gestapo*. But after everyone assured him Werner wouldn't be told of Mertens' identity, he agreed to have the youngster join.

Schneider brought him into the room. He was a very pleasant looking lad, with light brown hair, blue eyes, and a narrow wispy mustache. Mertens introduced himself as an art dealer from Berlin having business in Strasbourg, and during the ensuing conversation he learned a lot from this very knowledgeable, cooperative young man.

Werner's father was a Polish national, who died several years ago; however, his mother remarried and his stepfather, an officer of the German army, served in Russia; while Werner and his mother lived in Lahr, their home town. His stepfather was a patriotic Prussian with a family tradition over many generations, that sent their sons to serve in the army.

Werner sympathized with the Polish people. He learned to hate the Nazis for their atrocities against Poland, and now he couldn't wait to be liberated by the Americans. Mertens took a liking to Werner and after a warm farewell the group separated.

As agreed, the party met the next morning at the Colline apartment where four bicycles stood ready to carry Mme. Colline, Dr. Rasche, his daughter Erika and Mertens to the Vosges Mountains. Eduard Colline declined this trip, since he had to manage his factory, and wanted to make sure Banzer's antenna wiring fabrication was going ahead.

For Mertens the bike-trip to Still was fascinating. The weather in early September was perfect and the cloudless sky spread a blue dome over the idyllic scenery. The four travelers followed a narrow asphalt road west, leading through orchards of standing fruit trees at the right and left of the road shoulders. Acres of apple, cherry, pear and even mulberry trees lay before them. Surprisingly, nobody had bothered to start a harvest; especially now during days of strict food rationing. Every so often the four stopped pedaling to gorge themselves on these delicious fruit.

Occasionally they passed through small villages. They saw stork nests made of twigs, hovering on church steeples and roofs. It was like pictures from a storybook. This was Alsace at its most charming quaintness. Mme. Colline related, that even King Louis the Fourteenth was taken by the landscape, for as his carriage entered Alsace he exclaimed: 'What a wonderful garden the Lord has created here!"

Mertens completely forgot the seriousness of his mission, because of the peaceful setting and Erika's adorable company. Her beauty aroused him again and again. He observed every part and movement of her. The tight-fitting, gray, long pants tucked into a pair of short, shiny, brown boots. Her bright red woolen sweater further complimented an incredible female shape. Then, the very attractive face of classical lines, together with the shoulder-length light blond hair, made her look like a Hollywood movie goddess. The way she glanced at him, made it clear to Mertens that she seemed to be attracted to him as well. He couldn't get over Erika's marvelous sense of humor. With an insinuating twinkle in her eyes; she could come up with some very funny remarks. All of this made her companionship absolutely exciting. Finally, in German he said to her:

"Erika, do you remember the Christmas party at the *Kaiser Wilhelm Institut*? We were a few years younger. But I never have forgotten the first time I saw you."

She replied: "Your talk makes me blush and now you appear here in Strasbourg, somewhat like the white knight on a horse, trying to save us. Anyway I am also very happy to see you again and of course I still remember the party. You wore a sort of navy jacket with gray slacks. On the jacket you displayed a round emblem of your university. I thought you looked so .... cute!"

"Yes my mother also thought I am cute," he replied. Both broke out in cheerful laughter.

Then, he asked her about her childhood and life in Berlin. She became very serious and somewhat agitated, while she told of her past: "I was only twelve years old when I lost my mother through a mysterious car accident. The driver in a black Mercedes hit and killed *Mutti* instantly. She was trying to cross the street. The driver and car disappeared from the scene of the accident without a trace. The police never found the evil perpetrator. My father believes this was no accident, but the Nazis were responsible, giving Papa the possibility of continuing his work at the *Kaiser Wilhelm Institut*. Employees of that institution were forbidden to have either Jewish ancestry, or Jewish wives. *Mutti* was racially Semitic but was a baptized Protestant!"

Mertens interrupted: "These crazy racial ideas show a sort of inferiority complex by the Nazis and emphasize especially Hitler's insanity."

Erika continued: "As a result of this horrible accident, I became closely devoted to my father, who tried his best to raise me. At school I had problems; because of my half-Jewish status. It was at the Lyceum or high school, where the poisonous Nazi propaganda had affected my classmates, and my school attendance became very unpleasant. Father decided to enroll me in a private school, not far from home, where nobody knew of my Nazi problem!"

Mertens was astonished that the Nazis had such a crazy racial doctrine. "Erika, tell me more about your background!"

Encouraged by his interest she explained further: "All of these experiences caused me to hate the Nazis with an angry passion. Living as a second-class citizen in Germany is not acceptable and there will be a day, when I will leave these mad people. I shall begin a new life in a new and free country. I have read much about the United States and my favorite authors are Jack London, Sinclair Lewis, Upton Sinclair, Zane Gray, and many others. I have chosen the United States of America for my future destination. It sort of represents a Shangri-La of freedom and should help to reinstate my honor!"

Her words and her idealistic attitude touched Mertens.

Then he started to tell her of his childhood and life in Chicago. She had many questions and he was eager to answer everything she asked of him. Erika, being interested in American Jazz, loved to hear his stories about the big bands and names like Ella Fitzgerald, Harry James, Duke Ellington, Louis Armstrong etc. Some of her favorite Jazz numbers were the crazy 'Tiger Rag', 'Time Signal', 'Caravan' and many other tunes. Mertens was astonished, that she knew so much about this music. She explained that many youngsters in Berlin were fans of American Jazz, and listened secretly to the London BBC broadcasts from where they could easily receive transmissions, even on AM wave length.

After a trip of about forty kilometers, they reached Still, a quaint little village marked by a German-style church and steeple with a weather-vane rooster on its top. The streets were paved with cobblestones, and many farmyards had a tall manure pile heaped in their center; a place from which the yard rooster could inspect his domain.

Directly west of the village, the Vosges Mountain forest began. A small farm, tucked away in isolation, was their destination. The place belonged to Jacques and Maria Dubois, whose oldest son had been drafted into the German army serving somewhere in Russia. Their fourteen-year-old daughter and twelve-year-old son lived with them.

After everyone was introduced, the hosts invited their guests to a sumptuous lunch, consisting of a vegetable soup, roast lamb, sautéed potatoes, cauliflower, red wine, an assortment of cheese and finally a custard dessert. The meal was a miracle at this time of food rationing and everybody was in awe, complimenting their hosts for this generous spread. Afterwards, the three men sat in a corner of the living room, where a glass of cognac and a cigar provided a suitable finale to this unexpected feast; while the women chatted together, preparing the ersatz-coffee.

Mertens explained to Jacques the purpose of their visit, namely, obtaining new identities for Dr. Rasche and his daughter. He also inquired if the requested radio-transmitter had arrived from the American lines.

''Yes, last night," Jacques said. A young Frenchman came to the farm and brought the transmitter, but he left immediately, to return to the allied lines. Before leaving he told me it was one of his most difficult assignments. He almost got caught because the German SS-patrols used shepherd dogs in the forest. The front line recently stagnated, giving the Germans time to arrange for stricter defense line controls."

Mertens told the farmer about Karl Banzer's assignment. "Now that we have the transmitter, Karl should visit you to complete his responsibility," he said. "Jacques, you must contact the carpenter, mentioned by Colline, who would make the tree-trunk receptacle, including the lid. You should arrange for a wagon with two horses, to transport the decoy lumber with its radio through the forest."

"Tell Banzer to visit me in five days, as by then I should have everything ready for him!" Jacques said.

"Now let's take a walk to my print shop and begin with the documentation."

Walking through the backyard, a distinct rumbling of artillery fire could be heard sounding from the west, indicating that the war zone was not far away.

Except for Maria Dubois and Lucie Colline, they all walked to an adjacent barn building, the inside was completely filled to the rafters and beyond with bales of hay. It looked as if no printing facilities could possibly be in this location. However, when Jacques pulled three bales away from a corner area, a door led them into a good size workshop. Moving the bales back into place, they switched on a light and entered the fairly large room. It contained various printing machines and photography equipment. 'What an astonishing set-up", thought Mertens. Most of the time was spent taking down the relevant information and producing the necessary passport photographs for the Professor and his daughter.

Afterwards, Jacques said: "Tomorrow afternoon I should have all the documents ready for you."

Back at the farmhouse, the Professor and Mertens started a discussion about the hydrogen peroxide propellant. It soon became obvious to Mertens, that although the Professor was an outstanding chemical specialist in this field, he was no engineer. He could, however, provide Mertens with the most detailed materials compositions, its dangers, and the measures necessary to control the chemical. He estimated how this powerful chemical could be applied. For him to understand its potential, he needed to know the kinetics, the yield and given quantities of this chemical in conjunction with conventional fuel oil. This in turn would tell him how much energy could be produced and how much oxygen would be available for a typical U-Boat crew. The Professor was happy to provide him with these data.

From the information received Mertens ran a few calculations and extrapolated various sizes of steam-driven engines, starting with the smallest size at about 3,000 HP, ranging to the largest output of about 15,000 HP. The results were startling, especially when he calculated the large versions. He immediately realized the enormous energy available affecting the propeller torque, rpm output, and speed of such a submarine.

CHAPTER 8

Early September 1944, Karl Banzer and Pierre Sabon, the local chimney sweep from the village, drove the provided horse drawn wagon, loaded with tree trunks, into the Vosges Mountains. The weather on this morning continued to be beautiful with a cloudless sky. The narrow lumber trail brought them deeper into the wild forest of tall elm trees, beech, oak and aspens. The tallest were the elms, which were covered by high climbing ivy.

In contrast to a German forest, the ground was covered with natural underbrush.

The radio transmitter was stored inside the tree trunk cavity. The antenna wires had been fabricated, and then stored inside three metal milk containers, which were placed on the milk-truck for its routine trip to the country. Eventually Banzer and the pails arrived in Still at the Dubois farm. Jacques Dubois kept his promise and provided the two horse drawn wagon, loaded with timber. The radio transmitter had already been placed inside the hidden tree trunk cavity.

Three locations within the Vosges Mountains were chosen for the antenna installations, the first and immediate one, was just a few miles south of *Sarrebourg*. After driving through various roads in the forest, they found an ideal spot on a high elevation, where there was not a soul in sight.

Pierre jumped from the driver's seat, removed one of the milk pails, then attached a pair of climbing-irons to his feet and a safety belt around his waist. Karl opened the pail and fed the antenna-wire to Pierre as he started to scale the elm tree. After Pierre had reached the top he fastened the antenna. Subsequently Karl took his end of the stretched out wire, walking it round and round the tree. The ivy-leaf camouflage worked well, and Karl admired Pierre's skill and perfect blending of the wire with the rest of the natural ivy growth. The low end of this antenna wire was long enough to be stuck into the ground, making it appear like it grew out of the soil. Assured of their privacy, they opened the inserted lid from the bottom of the tree trunk, which rested on the wagon, and then removed the radio transmitter. Karl connected the antenna wire, slipping on the earphones as the battery warmed up the radio tubes. After he had selected the previously-agreed transmission wavelength, and reached the exact prearranged time on his watch, he spoke into the microphone. They hoped that by speaking English, any listening Germans would get confused, believing an American military patrol had penetrated the area.

"Bravo squad calling! Bravo squad calling. over!" Banzer released the transmission button and waited for an answer from Zürich. To his delight, he received a prompt answer.

"This is Richard squad, this is Richard squad! We can hear you loud and clear, go ahead!"

'We have made contact with phase four"! Karl replied. 'We are preparing for phase five!" This meant that they met the professor and his daughter. Phase five referred to their planned escape to Switzerland.

''Timing for phase five will be given at interval three or later!" Karl continued. This was their code for saying the next transmission would take place at the exact same time in three days and or intervals of additional three-day periods. Karl then concluded this first radio contact.

After the equipment was returned inside the tree trunk and leaving the first antenna installed on the elm tree, Karl and Pierre headed south somewhat west of Lutzelhaus, a distance of about 20 kilometers. They found another suitable tree, where they repeated the second antenna installation without any problems. This time Karl directed his transmission toward Paris. He repeated his first call identification procedure and to his delight received an answer. ''This is Bernard, this is Bernard, go ahead I hear you loud and clear." Bernard no longer needed to hide his radio equipment and instead had installed a most up-to date radio transmitter inside his St.Cloud office. Again, Karl relayed his message about the Professor.

It was getting late in the day and after the equipment was safely stored, they drove their horses and wagon to a nearby village where Pierre's brother lived. Here they rested and stayed over night.

At about five a.m., they rose and continued their trip to the last destination, an area near the old ruins of 'Burg Niedeck', an ancient castle, reputed to have once been inhabited by giants. The road climbed steadily inside the forest, winding its way through a peaceful vacation area. Finally, having reached the high point they could see the old ruins looming between 'the trees. It was almost six a.m.; the sky was dark and gray. Pierre and Karl were all alone, or at least so they thought.

This time Karl wanted to climb the tree and try his skill with the climbing irons. Pierre fed the antenna out of the last milk pail. Quickly Karl climbed to the top of the elm tree, and Pierre had just completed the ground insertion of the antenna, when suddenly, 'without warning, a motorized assembly of SS-soldiers drove up. Pierre was stunned to have been caught so completely off guard. The dense forest had absorbed the motor noise.

"What are you doing here!" shouted the *Schaar-Führer* (NCO), pointing his stiff baton at Pierre.

"I'm collecting wood and I'm looking for mushrooms!" replied Pierre.

"What do you have in these milk pails?"

"Nothing yet I haven't found any mushrooms!" was the reply.

"Open the pails!" was the command. After the NCO had verified the pails were empty he shouted: "Get the hell out of here; we have some urgent business and we don't want you around, *mach' schnell verschwinde*, scram!"

Pierre, leaving Karl high up in the tree, mounted the driver's seat and prodded the horses to get moving. He was apprehensive and hoped he wouldn't be followed. Whipping the reins onto the mare's back and shouting *"huuh"* he got going. The wagon started downwards on the slope toward the next village. Cranking the brake rod frequently to make sure the vehicle didn't run into the horses he finally reached the village.

Pierre estimated that Karl, on his own, would find his way back to Still.

In the meantime Karl, from his lofty spot up in the tree, heard shouting and the arrival of SS-soldiers. He knew the leaves were shielding him from visibility below; however, he could still observe most of the activities on the ground. He was in a cold sweat wondering if Pierre would get away safely from the 'SS'.·He was much relieved to see him disappear, unmolested in the direction of the next village. Clinging tightly to his swaying perch, he tried to observe what the SS-soldiers were up to.

Approximately one hundred yards away from Karl's tree was a clearing, where two canvas-covered trucks and one sidecar motorcycle stopped. The SS-Sergeant jumped from the motorcycle and shouted toward one of the trucks. Karl recognized immediately his Paris acquaintance, Horst Stolz.

About twenty SS-soldiers with a collection of shovels jumped from the first truck, opened the back of the second, and with rifle butts prodded the occupants into the open.

Karl, a seasoned intelligence agent was shocked. A group of twelve scraggly men in striped pajama type clothing and shaved heads stood together. Each one of them had a shovel thrust roughly at them and they were ordered to dig deep individual holes. If they didn't work fast enough they were beaten and kicked. Finally, the SS-Men were satisfied and stood six of their prisoners in front of their holes. Karl nearly gagged when he realized the next step. The prisoners shouted *"Vive la France!"* Then the poor victims were shot and dropped into their graves.

The remaining six prisoners were ordered to fill the dirt back into the graves and the operation repeated itself with another three victims being murdered. Again the remaining three prisoners closed the pits. Finally, two more victims were shot and the last prisoner had to cover the two holes. Before the last ashen-faced man was executed he also proudly shouted *"Vive la France!"* and the drama ended with two of the SS-soldiers filling the final grave.

As he wrapped trembling arms and legs around the unsteady tree-trunk, fighting nausea, Karl swore he would seek revenge on this monstrous NCO.

He knew his identity and he also recognized the familiar license plate on the back of his motorcycle, reading SS-A 234. Karl couldn't miss the arrogant manner in which the NCO carried the baton under his arm.

Late in the evening, Karl returned to the Dubois farm where Pierre nervously waited. They reasoned that the brutal murder of twelve Frenchmen was undoubtedly in response to *Standarten Führer* Heiner Kempe's orders. The message to the concentration camp in Schirmek must have been relayed by telephone. The execution place was not far away from the concentration camp. Stolz, being in charge of the execution presented enough proof that Heiner Kempe had given the orders. Seemingly, only twelve hostages must have been available; hopefully the other missing eight would never be found. Karl reminded Pierre to make sure he memorized the license number of the SS-motorcycle, and to advise the French underground for immediate action to locate and execute the SS murderers.

CHAPTER 9

Karl Banzer hitchhiking finally made his way back to the hotel in Strasbourg, where he met Peter Mertens, who was anxiously awaiting him. Mertens was outraged by the news of the murder of twelve prisoners. He decided to inform the French underground immediately, requesting them to execute the two *Gestapo* thugs in the hotel in retaliation.

Karl also reported his successful contact via the tree radio with Richard Johnson in Zürich and Bernard in St.Cloud, adding that Richard would be notified again in three days at the exact same time, when the Professor would arrive in Switzerland.

Madame Abel was home, as were the two Frenchmen, the German boy, and of course the Professor with his daughter. Having greeted everyone, Mertens and Karl asked for a private word with Lieutenant Pierre Schneider and Captain Didier de Roos, who after learning of the massacre at Burg Niedeck, became very grim. "Leave this matter to us; we'll take care of it!" said de Roos.

Mertens looked at de Roos, recognizing a deadly determination that made it clear that no further discussion was needed.

"Let me talk to Werner Garewski", said Mertens: 'We'll have to tell him not to bother keeping track of the two *Gestapo* agents. In fact, I don't want him to be involved with us at all. He's too young!"

As requested, Werner joined the group. "Werner, we believe it is better for you not to visit the hotel anymore; it is too dangerous," Mertens said. "If they put you under pressure, even without your wanting to expose yourself, you could easily reveal the existence of our little group. Please tell me what your plans are and what you intend to do?"

Werner looked disappointedly at Mertens, for he sincerely wanted to contribute to the downfall of the Nazi terror, but he was wise enough to understand Mertens' misgivings.

"I understand your concern and, of course, I will follow your advice," he said. As to my own plans, you know I have been waiting anxiously for the Americans to arrive, but they don't seem to be in a hurry. I'm getting so restless, I'm ready to explode. I'm seriously thinking of leaving Strasbourg, to visit my Uncle Kurt Huntzinger, who is a schoolteacher in Stetten, a small village right on the Swiss border. Maybe I'll take a chance and escape to Switzerland."

Mertens became most interested when he heard these ideas; it might be the perfect opportunity to get the Professor over the border into Switzerland.

'What kind of a man is this uncle of yours?" He asked.

"He is a Nazi party member; therefore I could never tell him about my real feelings and escape plans."

Peter knew instantly what should be done. A carefully established plan had to be worked out to take advantage of this great opportunity. He asked Werner to take a walk with him, and to discuss a plan involving both Werner and his uncle in Stetten.

After passing the municipal theater and walking for about twenty more minutes, Peter and Werner reached the Orangerie Park preserve, where they found a quiet, empty bench. Above and around them the horse chestnut tree leaves were turning yellow, and some of them already covered the ground, while the spicy fragrance of fall filled the air.

Carefully, observing Werner's face, Mertens explained the importance of Professor Rasche's knowledge of the German secret weapons program. 'Werner, here's your chance to contribute to the downfall of the Nazis.

Would you be prepared to let the Professor and me come with you and meet your uncle?"

"I'll be more than pleased to help where and whenever I can," the youngster said with a smile.

"In case we must provide the name of the person we are going to visit, your Uncle's party membership should be helpful...At the expected check points, police inquiries etc., we must have a perfect cover story explaining the importance of our trip. The documents and identity papers of the Professor, as well as my own credentials, will name us as art experts and curators.

Werner's eyes widened with excitement. He would love to help.

Finally, he got his chance to get even with these horrible Nazis. The idea dawned on him—clearly he would be working for the Allies. Mertens was probably an agent for the Americans, he thought.

"Our fictional cover story," Mertens continued, "will be that we are seeking safe storage for salvaged art treasures, away from the war zone. Nobody could find a better place, far and remote from bomb attacks, than right on the Swiss border. Therefore, the fictitious purpose of our trip will be to find a suitable barn or something similar as a storage area. It should be easy to sell anyone on this story, including your uncle."

Werner became even more enthusiastic after he had heard and understood this plan and offered his fullest cooperation.

"Could we get started in two days, early in the morning?" Mertens asked.

"No problem at all, I'll be ready to go, but on the way to Switzerland, I would like to stop at my hometown, Lahr, and say goodbye to my mother. Of course, I will tell her only about our cover story."

Mertens was satisfied with the arrangement, and believed he could set this assignment phase finally in motion.

After the two returned to the apartment, Mertens went to speak with the Professor and his daughter, explaining his sudden decision to begin the escape to Switzerland in two days. At first the Professor hesitated, being very concerned for his daughter's safety, but Mertens convinced the Professor that Erika should stay behind because of the dangerous border crossing. Reluctantly the scientist agreed to cooperate. Mertens asked Erika to wait for his return after he had guided her father safely into Switzerland.

Back in the hotel Mertens met Karl Banzer and brought him up to date with the events. He asked him to contact Zürich by transmitter, advising that in three or possibly four days, the Professor would arrive in Switzerland, at an area near the Swiss town of Eglisau. Then Mertens went to see Eduard and Lucie Colline in the rue de Bouxville, telling them of the latest developments, including the pending departure in two days. Mertens had become quite attached to this friendly couple. He promised to return to Strasbourg within a few days.

It was nearly noon and he was on his way back to the hotel when, having reached Hallen Platz, he suddenly heard the ominous drone of airplane engines. The ground started vibrating from the mechanical intensity of the aircraft flying in tight formation at an extremely low altitude. In no time at all the planes were directly overhead and Mertens heard the nerve-shattering screech of the first bombs heading in his direction.

Instinctively he threw himself flat on the sidewalk's pavement. Suddenly, an inferno of explosions erupted; the shock of the enormous air pressure almost ripped off his clothes. Less than hundred yards in front of him he saw a bomb slam into the ground, no explosion followed; it was either a dud or a timed device. Most of the bombs seemed to be falling near the railway station, and in the direction of the old town, where many unsuspecting people, walking outside in the beautiful sunshine, were caught by shocking surprise. The civilian casualty rate could be very high, Mertens thought, as he hugged the ground.

As sudden as the attack started, it was over; the American bombers disappeared. Mertens staggered to his feet, trying to clean the dust from his clothes. In a daze, he found himself returning to Colline's apartment, where the entire facade of the building had been damaged. Right in front of the complex a bomb had detonated, throwing the cobblestones high up into the air landing on the roof, some of them remaining lodged in the gutter; the front door was torn off its hinges and was leaning against its frame. Mertens rushed upstairs to his friend's apartment, where he found Eduard standing, still dazed and motionless, at the open entrance door.

"Are you all right?" Mertens shouted.

"Yes, Lucie is also unharmed. She's looking for our cat, which has disappeared!"

The apartment was in a shambles, with most of the doors ripped out of their frames. Fortunately, when the bombs exploded the couple had started their lunch in the kitchen, located at the back of the apartment. It was only in the front rooms, where all windows were shattered and extensive damage occurred. The scared, but unharmed cat, was found under the stove.

'Why have the Americans bombed our beautiful city? Don't they know we are their friends and we are French?"

"The city is crowded with Nazi officials and I am sure our military command had reasons to order this surprise raid," Mertens answered. The Germans didn't even have a chance to sound a warning alarm."

The Collines soon calmed down, and congratulated each other on having survived this terrible ordeal.

After Mertens had helped the couple clean the worst of the mess in their apartment, he said farewell. On the way back, he was worried about the unexploded bomb lying buried under the Hallen Platz, but everything remained quiet. The square in front of his hotel had been hit directly by a couple of bombs, just a few feet from the hotel entrance, whose lobby windows had been completely shattered and the facade severely damaged. Walking into the lobby he saw a line of about ten bodies lying on the floor, covered by blankets. One body drew his immediate attention; because of the large feet pointing forty five-degree angled sideways. Mertens walked over to the corpse and lifted the blanket from its head.

"It's Paul Krause. I wonder what happened to Hochstetter, maybe he also got killed?" he thought.

Arriving in his room he was glad he had left the hinged windows open; nothing was harmed.

He set out to check the apartment at the Ketten Gasse, to make sure that none of his people had been hurt. His emotions surprised him when he discovered that he was especially concerned for Erika. Approaching the front of the old structure, he saw there was not even a scratch to be seen. Relieved, Peter entered the building feeling a strong yearning to be with Erika. She affected him to such a degree, that he had to be cautious not to lose his pragmatic thinking capability.

"Hopefully, everyone stayed home during the raid and nobody got harmed." he thought to himself. He found everyone unharmed in the apartment. He told about his latest activities. Especially, Krause's death was good news for the two French officers.

De Roos promised to establish the whereabouts of the missing *Gestapo* agent, Hochstetter, and that inhuman Schaarführer, Horst Stolz: "If they are still alive they won't be for long!" he promised everyone.

The bombing raid convinced the Professor it would be a good idea to leave Strasbourg. But now, more than before, his only concern was for his daughter, Erika. Mertens explained that the purpose of the raid undoubtedly was to hit the many Nazis and French collaborators in town. He didn't think a second attack was imminent. The Professor and Werner agreed they would gather their personal effects and get ready for the journey to the Swiss border the next morning.

CHAPTER 10

After breakfast at the hotel, Karl Banzer decided to visit Mme. Abel's apartment to check if Peter, the Professor, and Werner Garewski had left for their Swiss border escape. Descending the granite steps of the hotel entrance into the Rote Platz, he noticed a parked motorcycle with sidecar standing at the curb. He recognized Horst Stolz's license plate number immediately, recalling the horrifying murder of innocent victims. He realized that Stolz must be somewhere in the hotel. He returned to the lobby to find out what he was up to.

Walking down the carpeted staircase from the first floor level into the lobby he saw *Gestapo* agent Ernst Hochstetter and Horst Stolz approaching. Both were in deep discussion. Stolz in a restless, aggressive mood tapped his baton against his leg. They noticed Banzer approaching and exchanged their usual *Heil Hitler* salutes.

'What are you up to today?" Karl asked. "I noticed your motorcycle parked outside. Are you planning a ride through the Vosges Mountains on this beautiful fall day?"

Stolz shifted his motion and slapped his baton into his left palm. "Are you crazy to suggest that I would go on a pleasure trip, while this plutocratic-Jewish war is going on!" He snorted. "We have important business to attend to and we are on the way to Schirmeck, where we'll sort out a few more French bastards in retaliation for the murder of our comrades in Paris. So don't give me this shit about a beautiful fall day."

Stalking off in obvious disgust, Stolz and Hochstetter left Banzer and walked swiftly to the hotel exit. Shortly after, Banzer heard the noisy sound of the cycle engine and saw them driving across the Rote Platz, where they disappeared around a street corner.

Banzer was frantic. He had to find a way to prevent another atrocity.

He rushed swiftly to the nearest telephone booth and called Eduard Colline who, fortunately, was in his office in Schiltigheim.

"Eduard! Here is Banzer! Stolz is on his way to Schirmeck with Hochstetter to take more hostages and murder them. We must stop them, if possible, before it's too late! They just left here by motorcycle, can you think of anything we can do?"

"You said Stolz left a few minutes ago?" Colline exclaimed after a moment of silence. "That means he should arrive in Schirmeck about an hour from now, around ten o'clock. One thing faster than a motorcycle is the telephone, which I'll use immediately. In the meantime, you visit Mme. Abel, reporting the situation to our friends. Once I am finished with my end of this affair, I'll contact you at Mme. Abel's place. I'll see you shortly, good bye!"

Banzer knew he had done his best considering the circumstances, and made his way to Mme. Abel's apartment.

In the meantime, Colline called Jacques Dubois in Still. He explained the latest news, imploring him to come up with some timely counter-action. He estimated he had about forty minutes to spare, before the motorcycle arrived in his area.

Dubois called Pierre Sabon immediately after talking to Colline, who agreed to meet him immediately, and within ten minutes both were on their way to the highway intersection leading to Schirmeck, reaching it within fifteen minutes on their bicycles. Strapped to Dubois' rear luggage rack was a carton of plastic explosives, part of which he had held hidden in his hay barn for future underground organization needs.

Having reached the highway, Dubois placed the worn and dirty brown carton close to the edge of the Schirmeck road, and hastily inserted an electric detonator fuse. Then he unwound an ignition wire stretching it about hundred feet away, through a grassy ditch to a point where both men could find protection behind a massive elm tree. In back of them was a tall cornfield where the two had hastily hidden their bikes.

Now they waited, hoping their rapidly put together plan would work, and that their timing would be accurate. Dubois held the electric detonator switch in his hand, trying to estimate the exact second he would have to use it.

They didn't have to wait long, for within five minutes they could clearly hear a clattering noisy vehicle approaching from the east. Yes, it was a motorcycle with sidecar. The passenger wore civilian clothes and the driver the uniform of an SS-man.

Stolz keeping his eyes on the road saw the carton and cursed: "These French pigs throw their garbage even on the highways!"

At the moment the vehicle was about to reach the carton, Dubois got ready to close the electric circuit. The two men crouched behind their elm tree before they could check the expected results.

When Stolz was about to pass the carton, an earsplitting detonation lifted the motorcycle clear off the road, blasting Stolz out of his saddle, somersaulting him into the ditch on the other side of the road.

Numbed by the explosion, shaken up and with his ears ringing, Stolz tried to get hold of himself. He knew the enemy couldn't be far away. He had to escape from this present danger; otherwise, someone would surely arrive and finish him off. Crawling on hands and knees, he moved out of the ditch into the protection of the adjacent cornfield, surprised that apparently no vital organs had been seriously hurt. He found he could stand easily, which he did. Next he pulled his *Mauser* Pistol from its holsters and swiftly disappeared into the safety of the concealing cornstalks.

Hochstetter was not so lucky; the blast had hit the sidecar first and overturned the motorcycle, with him still securely inside the cab. The machine landed upside down, hard on the asphalt pavement, smashing Hochstetler's head on the road's surface, simultaneously crushing him and breaking his neck.

The wheels of the vehicle were still turning when Dubois and Sabon emerged from behind the elm tree. Beneath the sidecar they could see a pool of blood spreading over the road. They knew instantly the *Gestapo* man was dead; but where was Stolz? Obviously he must have gotten away into the adjacent cornfield. It was much too dangerous to go after him. Therefore, Dubois and Sabon decided to roll up the remainder of the detonation wire, take the electronic gear, and immediately ride their bikes away from this scene of deadly retribution.

Inside the cornfield, Stolz finally stopped walking, and listened carefully to determine if he could hear voices or any type of human activity in his vicinity. He sensed he was not badly hurt, but his bones and muscles were starting to ache all over. "What happened to Hochstetter?" was his immediate thought. He decided to return to the road and survey the situation from a hidden position.

Backtracking, he quickly returned to the edge of the cornfield, peering out from between the stalks, but could observe no trace of human activity. To be sure, he waited a few more minutes, before eventually deciding to walk over to the motorcycle. Just then he heard the sound of a truck, and saw it approach from the west. After it stopped, two men hastily left their cab and offered help.

It was obvious, Hochstetter was beyond any help, prompting Stolz to curse and swear swift revenge. "These cowardly French bandits must be found and brought to their deserved punishment!" he shouted."

With the help of the two truck drivers, they were able to right the motorcycle by lifting the sidecar and turning it one hundred eighty degrees over to the other side. Then they removed Hochstetter's body from the cab, and placed it in the back of the empty truck. Stolz ordered the truck drivers to bring the body to a morgue in Strasbourg. As a final act, they moved the motorcycle to the roadside and out of the traffic lane, and then they all got aboard the truck and headed for the city.

Back in Strasbourg, Stolz called Kempe in Berlin and reported his latest encounter with the French underground. In spite of Kempe's raging anger, he ordered Stolz to forget the events in Strasbourg and return to Berlin as soon as his motorcycle was working again, because he needed him urgently for a new assignment.

Dubois and Sabon made it back to Still and called Colline's office in Schiltigheim, only to learn he had left for the day. On hearing this, Dubois decided to bicycle into town, where he arrived at Mme. Abel's apartment late in the afternoon. He reported the partial success of his mission, warning that Stolz was very likely alive, but mad as a hornet on a hot summer day.

Chapter 11

Early the following morning Peter Mertens, Werner Garewski, and Dr. Ernst Rasche, now known as Dr. Ernst Berger, art dealer, left Strasbourg traveling by streetcar towards the final stop on the line, the Rhine Bridge leading into Germany. The only luggage Mertens and Dr. Rasche carried was one soft leather briefcase each. Werner had a so-called *musette*, or a type of shoulder strap bag of cloth hanging on his side.

When they arrived at the Rhine Bridge it was still dark; the horizon beyond the river blushing with the first streaks of dawn. The bridge would lead the trio to the German town of Kehl, from where they could take the railroad toward the Swiss border.

The steam-driven, local train left Kehl with the three on board, as the first rays of sun started to bathe the landscape with its yellow light. Werner, who had taken this trip frequently, warned of the possibility of attack by Allied fighter planes. Their favorite targets were railroad trains.

Right and left of the track, as they traveled, was an expanse of meadows and corn fields, while the contours of the Black Forest loomed in blue-grayish colors from the east. The sky overhead promised another beautiful day, and one could almost forget the horrors of the war.

Werner was telling of his life in the beautiful province of Baden, how he went pheasant hunting, and how he loved to fish for salmon and pike in the backwaters of the Rhine River.

As the train moved, many pheasants could be seen from the windows, gorging themselves on the ripening com. Normal life had vanished; nobody had time to go hunting. During war the only prey is the human animal.

After numerous stops the train arrived in Lahr, without being bombed or strafed from the air. Mertens thought attacks would have been quite easy, because the Germans adhered to a precise on-the-minute train schedule; therefore, all a pilot needed was a railroad timetable and a watch to find easy targets.

At the Lahr train station, Mertens and the Professor agreed to wait for Werner, while he visited his mother. The next train to Lorrach, via the city of Freiburg, was scheduled to leave at 14:02. The delay gave Mertens and the Professor ample time to enter the station restaurant, have lunch and talk. Mertens was curious to learn more about Erika and the Professor's life in Germany. After some hesitation, the scientist slowly opened up and told his story.

"As a young man I studied physics and chemistry at Heidelberg University, and it was there I met my wife Luise. She came from a socially prominent Jewish family of Rhineland wine merchants. She studied art and music and was a most talented musician. After we graduated in 1914 and just before the outbreak of World War One, we were married. I was privileged to join the prestigious *Kaiser Wilhelm Institut* in Berlin-Dahlem, where I worked under Professor Fritz Haber, who during that period, discovered the ammonia synthesis from hydrogen together with the reduction of nitrogen, for which he received the Nobel Prize in 1918.

"Professor, I have visited the Institute a few times during my studies in Berlin and I am familiar with the organization," Mertens intejected.

"I was never drafted into the army," the Professor continued, "because my work on the development of poisonous gas had priority. But today, I am ashamed that I ever was involved in that horrible development. It was the time when Max Planck was running the research institute. Many famous scientists became my colleagues. Men like Albert Einstein, Nils Bohr, Otto Hahn, Lisa Meitner and many other outstanding personalities worked at the institute."

"In 1920 our daughter Erika, our only child, was born in Berlin and, being a bright girl, we gave her the best education available. We had a beautiful home in Berlin's Grunewald section, enjoying the incredible lifestyle offered by Berlin, then the world's cultural center."

"Professor, you have a wonderful daughter and I can only congratulate you, that you were able to bring her up as such a beautiful young lady," said Mertens. Looking almost embarrassed, the scientist continued his story.

"In 1933 the Nazis emerged, immediately driving out my Jewish colleagues and all Anti-Fascist thinking people. The Nazis demanded that I divorce my Jewish wife in order to keep my job, but I vehemently refused to do so. Like Prof. Haber, I also believed the Nazis wouldn't last and eventually Germany would recover its senses, but how wrong we were! Haber eventually fled to Switzerland, where he died in January 1934, of a broken heart, I believe.

"On the 30th of June 1939, five years after the day when the Nazis murdered Ernst Roehm and his gang of thugs, my wife Luise got killed by a hit-and-run driver, who never was found. Today I am sure the Nazis killed my wife to ensure that I would remain at the Kaiser *Wilhelm Institut.*"

At this point of his story Dr. Rasche almost broke down and it took him a while to recover his composure before continuing.

"In February of this year, one of the lab technicians, a Nazi party member, overheard a conversation I had with one of my assistants, whom I thought I could trust. In my frustrated rage against the Nazis, I said that Hitler was a criminal and a murderer, who would lead Germany to complete destruction, with the result that the whole world would hate every single German."

"Fortunately, the eaves-dropping Nazi lab technician told one of my friendly colleagues that he would report me to the *Gestapo*, and that I would be deported to a concentration camp. My friend warned me in time, and I had to flee Berlin with my daughter, where we were lucky enough to find shelter with Madame Abel in Strasbourg. Madame Abel had been a close friend of my wife Luise for many years, when they studied art at the University of Heidelberg."

Mertens was deeply touched by the Professor's story; especially since he cared so much for Erika. When the Professor began speaking again, he changed the subject and it was in a rather lecturing style he continued.

"There is one thing that really concerns me, and you, Mr. Mertens, being an engineer, you will understand why. At the *Kaiser Wilhelm Institut* a number of scientists worked on the splitting of atoms, and the results of their findings indicated that it is absolutely feasible to build a bomb which would be so powerful that one single device could wipe out a city like London. The material needed for this development would be uranium, because we learned the ore emits neutrons, which in turn would be the basis for a self-sustaining, "Prompt Fission" process; in other words a nuclear explosion. Our scientists, under the leadership of Prof. Werner Heisenberg, have discovered that self sustaining fission could be achieved through the use of a moderator such as deuterium, more commonly known as, heavy water. The moderator would serve to slow down the emitted neutrons just to the right velocity in order to have them bounce against the nearest atoms. They would dislodge more neutrons, which in turn would continue a self-sustained chain reaction. All this knowledge has been acquired through precise calculations, while steps were taken to secure the necessary heavy water from Rjukan, in Norway.

Mertens was fascinated by these words, and it confirmed his suspicion that the Germans were working on nuclear physics in a most dangerous endeavor. He had to control himself not to interrupt the Professor's lecturing speech.

"Fortunately, in 1942 the plant in Rjukan was blown up by saboteurs and the production of deuterium was delayed by two years. Last February the Norwegian operation was to be transferred to Germany, together with the remaining supply of heavy water. Again, saboteurs were successful by sinking the ferry carrying the heavy water tanks across a fjord. Now, my present fear is that someone will find a substitute moderator, making the need of deuterium altogether unnecessary."

Mertens, while at the Berlin University, had heard from other students about atomic experiments at the *Kaiser Wilhelm Institut.* As soon as he was able to return to Berlin, he planned to look into this matter with special care. He knew Hitler would be crazy enough to use this mass destruction weapon, together with his newly developed miracle rockets. As a result, he might still be able to win the war.

More than ever, Mertens was determined to get Dr. Rasche into Switzerland, so that his vital information could reach Richard Johnson and eventually the Washington high command.

"Do you know anyone at the *Kaiser Wilhelm Institut* whom I could contact, while in Berlin?" Mertens asked. "I mean a trustworthy person who also wants the Nazis defeated." For a moment the Professor looked uneasy, and Mertens noticed he was deep in thought.

"Yes, I am thinking of Dr. Herbert Hassenkamp, who works for Dr. Werner Heisenberg. He is a devout anti-Nazi, and you could reach him in Berlin's section called Dahlem; I am sure he is listed in the telephone book."

Werner returned to the railroad station. The train arrived punctually for the trip heading south. Finally, after they had left Lahr, the Black Forest Mountains became closer and closer to the railroad tracks. Sitting in their compartment and shortly after passing the town of Freiburg, the three travelers were approached by a *Gestapo* agent, who requested to see their identification papers. Mertens and Dr. Rasche, alias Dr. Berger, had no problem passing inspection, especially after they explained the purpose of their trip. But when the agent examined Werner's documents, he became agitated and shouted.

'Why aren't you in the army? Your *Wehrpass* (Military Passport) indicates you are eligible for service in the *Wehrmacht!* This matter has to be further investigated. Come and walk in front of me toward the head of the train."

Mertens was alarmed by this sudden development and told the Professor to stay in the compartment while he would follow the two. The train was fairly empty, since all the larger townships had been left behind in the north, and the Swiss border at Lorrach was about 30 kilometers away. With him following a short distance behind the *Gestapo* agent, and Werner just ahead of them, they reached the first car next to the locomotive. There Mertens saw a special compartment marked with letters in black on a white background: "*Wehrmachts Polizei*" (Military Police). Brown curtains covered the compartment windows.

Mertens' thoughts raced desperately to prevent the agent from reaching the army officials. Just when he decided to attack the Nazi policeman, the train erupted in a convulsive upheaval, with the crashing sounds of bullets penetrating the right side windows. The railroad car seemed to lose its balance and started to jump off its tracks. A deafening hissing sound pierced the eardrums, indicating the locomotive's boiler tubes had been hit, and after a number of thudding jerks the train finally came to a stop, with the first passenger car leaning over on its left side.

Everyone was lying stretched out on the floor. Mertens, the first to recover his senses, swiftly lunged forward on top of the *Gestapo* man, and with a judo blow, broke the agent's neck killing him instantly; then he recovered Werner's *Wehrpass* and helped him get up. The young man seemed unhurt, so they moved to the nearest exit door before the MPs in the compartment could scramble out of their trapped situation. The car, lying at an angle, made it easy to open the outside door, which swung open by its gravity. They finally found their way to the outside, standing near the train. Recovering for a few minutes, Mertens and Werner walked toward the back of the train, searching for the Professor.

Most of the passengers had fled the train, seeking shelter on the railroad embankment in case the air attack should continue. The Allied pilot, seeing the results of his attack, banked away and disappeared toward the west.

They found the Professor toward the rear of the wreck on the embankment, obviously unscathed. The assault had been concentrated on the locomotive at the head of the train. Mertens told the Professor the Gestapo agent had been killed in the attack, and he had been able to retrieve Werner's *Wehrpass*. Werner, hearing this explanation, couldn't help but produce a diabolic grin. He was indebted to

Mertens and was most grateful, that this dangerous situation had ended so well. He quickly returned his *Wehrpass* to his pocket.

Mertens decided they should walk over to the country highway, which ran along the railway line leading south toward Lorrach. He was thinking that walking would be better than waiting for some unknown makeshift transportation. Each had only a light load to carry, and he estimated the border town to be about twenty-five kilometers away. Fortunately, after only a few minutes walking, a Red Cross panel truck appeared heading south, and Mertens flagged him down. After hearing their story, the driver gave them a ride. Later in the afternoon they arrived with the truck in Lorrach, where the Swiss border point was clearly marked by a tethered barrage balloon floating in the sky.

To continue on to Griessen by train was risky, because of Werner's army *Wehrpass* predicament; however Mertens decided to take a calculated risk and make the trip. The three bought return trip tickets to Lorrach. The train was scheduled to leave at 18: 15 from Lorrach. They didn't have to wait long and Mertens chose, just behind the locomotive, an empty railway car with wooden benches. He had learned a few hours ago where to be on the lookout, and that *Gestapo* controls started from the rear.

Each railroad car was separated by an open, steel-plated platform with a handrail, making it possible to walk from car to car. Looking through the glass-paneled doors, a person could observe the activities of the other car compartments. Only a few passengers were on board, mostly mothers with children or elderly villagers. The train, clearly a local, stopped frequently at all small hamlets.

It was turning dark outside as the mighty Rhine River appeared on the right hand side of the train. The swift current and eddy streams of the river were shifting back and forth over the water's surface. The sight of brightly-lit villages across the river in peaceful Switzerland was unreal. In contrast, every street and home in Germany was strictly controlled by wartime black out.

Just before reaching the town of Tiengen, Mertens noticed a tall man in SS-uniform approaching them along the third car from the locomotive. He carried in his left hand an old fashioned, oil lantern. Most of the travelers seemed to know him for he chatted with a few of them as he moved along the aisle. As he approached the car next to our travelers, the train came to a stop in Tiengen. At a pre-arranged signal, the three men rapidly left their compartment, stepped onto the station platform, and walked quickly toward the end of the train, where they re-entered another train compartment. This situation could have become dangerous, but, luckily the SS-man didn't notice the fast exit of the three.

During the dark evening hours, they arrived in Griessen. Suspecting that a policeman might check the arriving travelers at the exit gate, Mertens suggested to remain on the station platform. They made themselves comfortable on one of the wooden benches.

The professor said. "Since we haven't eaten in a long time, we should find ourselves some food in town."

"That won't be necessary," Werner said, surprising them all. My mother gave me a number of sandwiches and a flask of lemonade, which we could take care of right now."

Mertens liked the idea, as this would help extend the waiting period before passing the exit gate.

After having eaten the pre-war type sandwiches of ham and sausage, the group made their way through the empty station toward the now unattended exit. Soon they walked in complete darkness toward the center of the small village. At the end of the main street they found a quaint two-story hotel, where after ringing a bell at the reception desk, an attractive middle-aged lady appeared. Mertens immediately took a liking to her, and in his most charming manner, asked if she had two rooms available for one night. After he had paid the friendly woman, she handed him two keys and pointed them upstairs, explaining each room had two single beds.

"Would it be possible to avoid any registration?" Mertens asked. 'We want to be undisturbed, as we are leaving early in the morning."

The woman smiled understandingly and replied. "Don't worry; I'll see to it that you'll sleep without being bothered."

Peter and Werner took one room while the Professor slept in the other.

They had a much-needed restful night and the bright early morning sunlight indicated another beautiful day was on its way. Thanking their hostess and without eating breakfast they set off on the road leading to the village of Stetten, where a yellow road-marker with black lettering indicated a distance of three kilometers. Soon they walked past a hidden guard-post. Fortunately, it was unoccupied. Obviously, it served as a night control point. Early in the morning the road was completely empty. Not a soul could be seen anywhere. On both sides of the path were bushes and trees. Stretched to the left and beyond the bushes were open fields, skirted by a forest. Rolling hills and a number of orchards, covered with fruit trees, shaped the landscape.

They noticed the sky turned drab, and a gray mist developed. Not a breath of air could be felt. After a half hour walk, the trio reached a downward curve, sloping into the village of Stetten, where Werner led them to the home of his widower uncle, the local school teacher and Nazi party member, Herr Kurt Huntzinger.

The house was a one-story brick home with a gray stucco facade and red tile roof. An air raid siren mounted its top. Herr Huntzinger knew the trio would arrive because Werner had made a telephone call from Lahr announcing their upcoming visit. Werner also had explained the reason and its purpose of seeking a shelter for art objects. The teacher invited the men to enter and make themselves comfortable in the living room. He was a lean and friendly man of about sixty.

Mertens explained the importance of his mission. Herr Huntzinger was quite impressed at being asked to aid in such a significant state affair. Of course he would be happy to find a suitable barn, or storage place for all of these exciting art treasures. At this point he offered lunch consisting of thin vegetable soup and some undefined cheese sandwiches ala chemical 'I.G.Farben.'

"Obviously, the local farmers didn't supply this man with wholesome food supplies." Mertens thought. "Maybe they didn't like his party membership?"

In the living room various school papers covered a desk. Mertens noticed a letter-size cardboard booklet with the title *Heimat-Atlas* or Home-Area Atlas. Casually, he asked the teacher if he could glance at this booklet to get an idea what the area looked like. Without hesitation the teacher handed him the book of maps. Soon Mertens found what he was looking for. He discovered a most detailed description of the village and its border with Switzerland. Just about every tree was marked. Feverishly he memorized the map. He now knew what to do and from where to start.

The teacher excused himself as he had some of his pupils' homework to review, so Mertens suggested that they take a short walk.

The mist had increased and the air stood oppressively still, ominous and heavy. Walking in the direction of the Swiss border, the road turned south to a border checkpoint about half a kilometer away. To be seen on the other side was a large Swiss flag with a red background and a white cross, hanging lifelessly from a tall white pole. Still nobody was anywhere in sight, so Mertens decided that this was the ideal opportunity to begin the escape operation.

First they climbed a grassy knoll covered by a number of fruit trees. Mertens ordered the other two to lie flat behind the trees, while he carefully scanned the field ahead. He could see the meadow was leading to the edge of a dark green forest. In the middle of the field, some fifty yards away, there was a camouflaged guard position, which Mertens saw was empty. As he looked toward the edge of the forest, suddenly almost unseen, a green uniformed border guard became visible, walking leisurely with his rifle shouldered. Fixing his eyes intently on the guard, Mertens watched this hostile figure moving slowly to the left, then completely out of sight.

He signaled his companions and they rose from their hidden positions and all ran as fast as possible toward the edge of the forest. Swiftly but silently they reached and penetrated the dark timberland of old beech, spruce, oak and pine trees. Mertens was worried about the Professor, but fortunately, he seemed perfectly able to keep up. The group moved from tree to tree working their way further into the woods. After descending and again climbing the cavity of a deep ravine they reached a dense nursery section of ten-foot high spruce trees. By this time they, were drenched in sweat, and breathing heavily from the tension and exertion.

They made their way slowly through the rough branches of the densely planted spruce until they reached the end of the section. From here the tall forest continued toward the Swiss border. Before leaving their well-sheltered area, Mertens ordered everyone to lie down, while he scanned the outer edge of the nursery. He then decided to proceed with the operation, penetrating further ahead and to their delight, discovered border-markers consisting of round, three-foot-high 'posts. The identification was painted in the red and white colors of Switzerland. It was here where the German forest ended, and a gentle grassy slope led up hill into Switzerland. Mertens signaled his partners to move forward and escape to freedom.

Elated and almost noisily jubilant they stepped over the border and began moving up the slope, with Mertens watching to be sure, the Professor and Werner made it all the way to safety. After they had gone about one hundred yards into Swiss territory a shot rang out, and Mertens saw the Professor fall to the ground, holding his upper leg in pain. At the same moment, a group of horse-mounted Swiss soldiers galloped over the knob of the hill, completely surrounding the refugees. Under the guidance of a Lieutenant, the soldiers inspected the nasty flesh wound of the Professor, and dressed it from their army supply kit. Then they produced a canvass blanket, placed the Professor on top and four men carried the victim about two hundred yards to the Swiss border guardhouse.

Even though his friends were now in safe hands, Mertens felt an angry rage take hold of him, over this brutal act. The Professor was apparently only wounded and hopefully would recover, but Mertens was furious that the Nazi border guard would shoot across into Swiss territory. He scanned the area from where he had heard the sound of the rifle shot, and quickly spotted a green-uniformed guard standing beside a thick oak tree, unaware that he was now being observed.

Mertens suddenly felt like he was back at Anzio Bay in close combat with the enemy. The guard was about one hundred yards away. Moving from tree to tree, he began to stalk the sniper, eventually working into a position behind him. Assuring himself that nobody else was nearby, he slowly approached the guard from behind and with a vicious hard chop to the back of his neck, he snuffed the guard's life.

As the dead man fell to the ground, Mertens suddenly heard the sound of a barking dog. Thirty yards away, in the knee-high growth of fern stalks stood a growling shepherd dog. His fangs were bared, and he was obviously trained to track fugitives, and sound the alarm. Mertens acted swiftly. Picking up the guard's rifle he checked to see if it was loaded, took aim and placed a bullet through the center of the animal's skull. From the guard's leather pouch, he removed several cartridge clips, containing five shots each, then moved quickly into the protective cover of the spruce tree section. As he pushed forward, he heard the shouts of other guards, who now had discovered the body of their comrade and the dead dog.

Mertens feverishly tried to recall the picture of the teacher's map while he stayed under cover, carefully moving north through the forest. The misty weather condition made it difficult to establish a clear direction, and the sun was no longer visible. In his Boy Scout days, and later ranger training, he had been taught, that the westerly weather side of trees is usually covered by a moss-type growth. Following these signs he cautiously moved ahead, in what he believed was a northerly direction. His main concern at this point was the existence of other shepherd dogs. However the weather, mist, and wind were in his favor. On several occasions, he came upon groups of grazing deer. They looked at him with their large dark eyes, seemingly aware that he also was a hunted prey. When he began to approach them, they moved further ahead. The smooth forest floor with its soft moss and pine needle cover allowed for a noiseless pace.

After about three hours of carefully navigating through the woodland, he again discovered a group of border posts, which he knew to be the area called *Schaffhausen Zipfel* (Schaffhausen, a land strip extending north over the Rhine River).

It was already early evening, so he decided to seek temporary-shelter within Switzerland, at least until the Germans had given up their search. Walking into Swiss territory, he was alone and undiscovered, but within a short distance, he had to cross a road leading westward again into Germany. Ahead, was more Swiss territory, where he saw an open field with a number of haystacks. Choosing one of these piles he buried the rifle and himself in the hay. He decided to stay hidden until darkness, to give himself more protection.

It was about eight o'clock in the evening when Mertens left the haystack and headed to the north of *Erzingen,* then he crossed again into Germany at a highway leading south-west. Being well within German territory, he pulled the trigger of the uncocked rifle. Now he could remove the bolt lock, and throw the weapon into a bush. Having walked another hundred yards, he discarded the bolt lock plus ammunition into a water-logged ditch. He crossed the road and walked along its embankment in the direction of Tiengen. Here he went to the railroad station where, using his return trip ticket, he began making his way to the final border town of Lorrach.

Knowing he had to get out of this immediate territory as soon as possible, he took the next available train from Lorrach at 22:43, leaving north for Kehl. Early in the following morning hours he arrived in Strasbourg, back at Madame Abel's apartment, where he awakened the landlady from her sleep.

The reunion was a happy one, especially after Erika appeared from her room, wrapped in her bathrobe. Mertens impulsively embraced her, saying that her father was safe in Switzerland. He had been shot in the leg by a fanatic border guard, wounding him while inside the neutral country; however, she should not worry. To his surprise Erika responded by giving him a warm lingering kiss with tears in her eyes. Soon the two French officers joined the group and Mertens told them about his trip. The two officers reported that Horst Stolz had left town with his newly repaired motorcycle. He probably would rejoin his boss Kempe somewhere in Germany.

Mertens informed Erika that he would like to leave Strasbourg the next day for Berlin. He asked her to join him because she could be of great help to him. Without hesitation, she agreed. Back at the hotel, he booked a room for one night, planning to leave the next morning with Erika for Berlin.

Picking up the hotel telephone, he reached Karl Banzer, who was overjoyed to hear from him, and they agreed to meet in his room. Karl opened a bottle of cognac and listened to Mertens' successful account of his mission. They agreed that the next day, Karl would go into the Vosges Mountains to radio the good news to Richard Johnson.

Karl reported that he had received word from Eduard Colline, who said the Americans were diverting war material to British General Montgomery's forces, so that he could start an all-out offensive in the Netherlands. That was the reason the Americans could not continue their successful offensive with General Patton's Third Army. Apparently, Montgomery was having a tough time with his largely Canadian airborne offensive, and was under pressure from England to eliminate the 'V-1' rocket launching stations in Holland. These missiles were causing great fear and damage on the home front, particularly the bombings of London.

Karl's partnership had come to an end. He was assigned to a new mission. Mertens thanked him for his great support and friendship. Let's meet again after the war is over.

CHAPTER 12

Early in October 1944, Mertens and Erika left Strasbourg, riding the streetcar to the Rhine Bridge. They traveled by rail to Appenweier, where they changed trains and made a direct connection to Berlin.

Although the compartments were fairly well-occupied, they found seats near the window, next to each other, along with four other passengers. Two suitcases, including Erika's cosmetic valise and Mertens briefcase, were stored in the overhead baggage racks. Mertens was delighted to have Erika all to himself, with nobody around to distract them. He suddenly felt responsible for her, which gave him a feeling of a delightful possession. Erika in turn was excited, and enormously grateful, that this handsome Adonis had taken such intensive interest in her. He was just the type she had dreamed 'about, a man of action, well-educated, who would be a marvelous companion.

Passing through Karlsruhe, they saw the enormous damage caused by repeated bombing raids, leaving many buildings still smoldering. Arriving later in Frankfurt, they entered a railroad terminus station, where the steam locomotive had to be changed. Again, they saw the devastating result of how a great metropolis became victim to incessant bombing raids. The formerly glassed-steel structure of the huge Frankfurt station was now an empty skeleton shell. Buildings were still on fire, and black smoke hovered over the once proudly-acclaimed cultural city, the birthplace of the great philosopher and poet, Johann Wolfgang von Goethe.

Now, the train became crowded. It was almost impossible to move along the narrow aisles leading to the individual compartments. Travelers with stacks of suitcases, including soldiers returning to their units with their gear, filled the corridors. Some of them stood or sat on bags and suitcases in these confining aisles.

Further north just before reaching Kassel, the train was delayed by numerous stops; because of on-going air raids ahead of them. It wasn't until later in the evening, when they finally started heading east toward Berlin.

In the darkened compartments, eerie black lights radiated a dim faint glow, causing unusually white gleams from some passengers' teeth. Mertens felt very satisfied at having the good fortune of sitting comfortably with Erika inside the train's warm compartment. He noticed a fragrance belonging to Erika next to him. It made his head dizzy. Yes, he had fallen in love with this young woman. She was so beautiful. Turning his head he admired her superior profile. He knew she was witty, charming and intellectually exactly, the type of woman he had always hoped to meet.

After a while and completely relaxed, Erika leaned her head against his shoulder. Attracted by an invisible magnet both embraced and shared a deep, passionate kiss. In this half darkness they were completely oblivious to the other passengers, who were mostly dozing and asleep anyway. Holding hands they silently continued their journey, every so often kissing and whispering intimacies to each other.

Prior to their arrival, Berlin had been bombed again. The train had to wait near the town of Brandenburg, before being allowed to enter the city. The journey ended early in the morning and inside the Anhalter railroad station.

To visit anybody unannounced at that time of day would be an imposition, but Mertens had no choice; he had to find a place for them to rest. The only contact he could think of was Herr Wilhelm Kuhn, who owned an art and antique gallery in the Keith Straße near Wittenberg Platz. This address was given to him in Washington as his Berlin contact point. Mertens knew it served also as Kuhn's residence. He decided to seek his help, early hour or not.

Walking a few of blocks from the railroad terminal brought them to a subway station with a direct connection to Wittenberg Platz. Shortly before reaching the *Gleisdreieck* station, the train left the underground area and climbed into the open onto an elevated system. From here they had a far-reaching view over the still-smoldering city. Mertens was horrified to see his beloved Berlin, in such a state of utter devastation. Erika, in spite of experiencing bomb raids earlier, was also shocked to see how much additional damage had been inflicted, since she had left, less than a year before.

Mertens, while at the Berlin University in 1938, had witnessed the November "Crystal Night." He couldn't help to compare his former experience with the present situation. The seeds planted in 1938 were now sprouting all over Germany.

It was seven o'clock in the morning, when they eventually stood in front of

Wilhelm Kuhn's gallery. The main show-room window was broken but covered by a solid rolled down shutter, while the former glass door, also shattered, was shielded with plywood panels. Berlin virtually had no windows left intact.

Ringing the doorbell, knocking at the plywood, and making a lot of noise finally brought results. The shop door was opened and inside stood a slender man of about thirty-five. He was of medium height and still fighting the sleep in his eyes. His pajama top was tucked inside old gray pants, while a pair of black leather slippers covered his bare feet.

In a hostile voice he grumbled: "What on earth do you want at this ungodly hour?" Mertens replied: "Are you Wilhelm Kuhn?"

The man answered: "Yes, that's me, so what do you want?"

"I would like to speak to Blondie!" he said.

For a moment there was silence; flabbergasted, the man tried to collect his thoughts, and then with a dawning smile he said: "Oh, you mean Jean Harlow? Excuse my appearance, but I haven't slept for a long time. These constant air raids prevent us from getting any decent rest. Please come in."

Once inside the store the man closed the door, turned on the two ornate ceiling lights, told the two visitors to leave their luggage in the front store, and led them back through a narrow passage to the living area. Mertens noticed Kuhn was limping. Toward the end of the passage, on their left were a number of doors, the first opening led into the living room. On entering Kuhn said: "Please sit down!"

After everyone was seated in plush armchairs, Mertens introduced Erika as Fraulein Berger, her new name. Then, very briefly he told Wilhelm Kuhn about their latest activities, including the escape of Professor Rasche into Switzerland.

Kuhn asked: "What can I do for you?"

"Erika and I need rest and a place to stay for a few hours."

Kuhn replied: "You can both stay here, Fraulein Berger in this room, sleeping on the sofa, while you can use my bedroom. In the meantime, I'll leave you alone, while asking some of my friends for a place for you to stay. As you can imagine, accommodations are a very big problem after so many apartment buildings and hotels have been destroyed. Anyway, I should be gone for about four hours. Then we must discuss our next moves and plans. Please help yourself to whatever food you find in the kitchen. I have a few foreign embassy friends who always supply me with valuable food items."

At that Wilhelm left, leaving Mertens and Erika to decide whether they needed food or rest first. Both slept for a couple of hours and then met in the living room.

Being alone and undisturbed gave them a chance to become better acquainted. Erika began to talk.

"When I was seventeen years old, I lost my mother through this mysterious car accident. My Papa hinted many times that this was no accident, but probably a murderous act by the Nazis. Anyway, I grew up to develop a great hatred toward Hitler and his vicious gang of killers."

"I certainly can understand your feelings, but don't let me interrupt," said Mertens.

"In high school we had a class called 'National Political Instructions,' from which all Jewish students were excluded. At that time nobody in class knew of my Jewish mother. I attended these stupid lectures, which were given by a political fanatic, Herr Dr. Fritz Heimwald. The man indoctrinated our group with the most vicious anti-Semitism. With chalk he drew the face of a so-called Jew on the blackboard. Of course the features included a big nose, a drooping lower lip and kinky hair. I noticed how this poisonous indoctrination took effect on my classmates."

"Philosophers once said, "give me your youth for indoctrination and I can change the future of a nation," commented Mertens.

"One day Dr. Heimwald brought along a curved tool, like a large size compass or tongs. He called a number of girls to his desk platform and measured their head features with the result of various racial interpretations. To my horror, he suddenly ordered me to appear in front of the class. After he was finished measuring my features, he firmly announced that I was an excellent example of an Aryan, Nordic woman. I didn't know if I should burst into hysterical laughter or cry."

"Erika, the idea of a pure Aryan race is only fantasy. Around the year one thousand, German knights colonized Berlin and all its eastern territories. They mingled happily with the native Slavic races and even today, you'll notice many Germans with Slavic names and features. Europe is quite a mixture of former tribes and migration waves. My country, the United States, is an enormous mixture of races and nationalities. The result is a synergy of inventiveness, entrepreneurship and boundless energy, unmatched anywhere in this world and totally disproving Hitler's ludicrous theories. "

Erika smiled approvingly at these remarks and continued:

"In September, 1935 the Nazis issued their 'Nuremberg Laws,' depriving Jews of their German citizenship and declaring Half-Jews as second rate Germans. Finally, in our school, all Jews and Half-Jews were excluded from political lectures. The rest of the students gathered every Monday morning around a flagpole in the schoolyard, singing Hitler Youth songs while raising the swastika flag. From then on my life in school became hell. I felt snubbed and was told, that I would never amount to anything because I was a Half-Jew. Fortunately, my Papa removed me from this school and enrolled me in a local private institution where nobody knew about my situation."

"More and more with a passion, I began to hate Nazi Germany. I started reading English and American literature, including the 'Outline of History,' by H.G. Wells. The free, democratic philosophy of these authors took hold of me and I swore to leave Germany as soon as I had the opportunity.

"In the meantime, I decided to actively fight and undermine the Nazi scourge. When I was about eighteen years old in 1938, I had my first opportunity. My parents had sent me to a ballroom dance school at the Kurfürstendamm, where I met a young man, named Helmut Anders. He was three years older than I, the only son of a widow, who lost her husband during World War One. Helmuth was a romantic person, who loved the fine arts and he taught me about French Impressionism and the German Avanguard movement. He hated the Nazis, with their crude art of robots, having small heads and super muscular physiques. He was a socialist and believed in Karl Marx, sympathizing With Communism. I fell in love with him and we became very close to each other."

"Erika, you are making me very jealous!"

"Helmuth is dead", she replied. "But let me continue. One night we strolled along Kurfürstendamm, when we stood in front of one of these hideous, windowpane enclosures, displaying the 'Stürmer' newspaper. The Stürmer is published by Julius Streicher, a fanatic Jew-Hater, who published almost pornographic descriptions of Jews, seducing innocent Aryan girls, usually jilting them in a pregnant condition. Standing in front of this display we were overcome by rage. The street was empty and no one in sight. Helmuth saw a loose brick at a nearby fence, picked it up and smashed the glass enclosure. Then we ripped the papers into pieces and we triumphantly left the scene of our accomplishment."

"What happened to Hellmuth?" 'The Nazis shot him. Actually he was drafted into the army and sent to the Russian front. In 1942, with the rank of corporal, he was sent to the town garrison of Schwerin, north of Berlin. Just before I was prepared to visit him, he went to a local pub, where he met two single girls. In his angry emotional state, coming from the Russian front he spouted off, declaring that the war was lost, and adding that Hitler was a monstrous criminal. The girls reported Helmuth to the *Gestapo* and the next day, dressed in his swimming trunks, he was shot and buried in the surrounding forest."

Mertens was stunned by this story and he became more determined than ever to successfully fulfill his mission. Before he could ask more questions Kuhn returned.

"The only place where I could find quarters is at the private home of a friend, living in Groß Glienecke, about 35 kilometers west outside town. He is willing to provide a single room for both of you," he stated.

Fortunately, it was Sunday and Mertens decided to call his old college classmate, Wolfgang Kopaneck. Wolfgang's mother answered the phone, and told him that her son had just moved to a place in the city. She didn't know exactly where, but she knew the telephone number.

Wolfgang was home and most surprised to hear from his old friend Peter. He immediately suggested they meet and gave an address near Halensee in a street not far from the Kurfürstendamm. Mertens advised Wilhelm they would return before evening and then decide what to do next. Mertens and Erika took the yellow double-decker bus from Wittenberg Platz to the intersection of Kurfürstendamm and Brandenburger Straße. Peter explained to Erika that Wolfgang was not too fond of the Nazis, and was a reliable friend. Probably, his main goal would be to survive the war unscathed. He presumably would also cheat the army or the Nazis if necessary, just to reach his own objective. Walking a few blocks south, they arrived at Wolfgang's home at Ravensberger Straße. It was a villa, enclosed by an iron picket fence with double-wing gates, leading to a garage. The property, surprisingly, was still untouched by air raids. Wolfgang answered the bell through an intercom and, after identifying his visitor, a buzz at the entry gate opened the door. Both friends embraced, and Peter introduced Erika Berger as his girl friend.

The villa was beautiful, prompting Peter to ask Wolfgang. "How on earth do you manage to live in such luxury?" Wolfgang replied. "Come inside, make yourself comfortable and I'll tell you all about it. By the way where did you find this gorgeous Lady?"

They walked through a spacious hallway, leading into a salon with a gas-fired hearth, around which a leather couch and a couple of leather armchairs were grouped. Peter noticed above a balcony walkway with railing, surrounding half the hall and adjacent salon. Various rooms could be reached from this balcony-like structure. One glassed skylight illuminated daylight to the area below. The walls were lined with floor-to-ceiling bookshelves, completely filled with fine-looking volumes. Peter thought it resembled· the home of a British nobleman.

After they sat down Wolfgang told his story.

"This house belongs to a courtesy uncle of mine, who is a bachelor, a German diplomat and history professor. He is deeply involved with the Third Reich administration and is presently outside the country on special assignment. He asked me if I would come to live in his home, while he was away, protecting his property from incendiary bombs. As you know, my mother's apartment is small, so I was more than pleased to accept his suggestion. But now, tell me about yourself and your beautiful girlfriend.

How come you are in Berlin? I always thought you are an American citizen."

Even though he trusted Wolfgang, Peter wanted to protect him with a concocted cover story in case someone should ask questions in the future.

"After our graduation in April, 1941, I went on a short visit to my parents in the States. In May, I decided to return to Germany and apply for citizenship. I believed Germany was the land of the future, and wanted to be part of this exciting development.

"Unfortunately, it was a mistake and now I regret the decision, but who could have foreseen that Hitler would start a war with Russia and even with the United States? At the moment, I am on special assignment to receive and transfer valuable art treasures from France into Germany. I have been living in Paris and Strasbourg, which is where I met Erika, the daughter of an art dealer friend of mine.

"I'm in Berlin to take care of some urgent business, probably lasting over the next three weeks, but I don't like the living quarters offered; they are too far out of town in Groß Glienecke."

"Why don't you move in with me? There are a number of bedrooms available, and it would be great to have you both staying here."

Peter was delighted to hear this. He decided to cancel Kuhn's offer and immediately accept the invitation.

"Wolfgang this is a beautiful idea, and I hope we don't cramp your style! If you agree, Erika and I will return to my art dealer friend in the Keith Straße , to pick up our luggage."

Wolfgang said: "I'll be expecting you in a couple of hours, OK?"

Back at Keith Straße, Wilhelm Kuhn was pleased to hear the news of having found these fantastic living quarters. Peter asked Erika to leave them alone because he and Kuhn had to talk business.

Erika understood and replied: "I shall take a little walk through the neighborhood."

Kuhn suggested visiting a nearby pub where both could talk in safety. You never know if my place is bugged."

The pub was a typical Berlin place, which during normal times, served excellent draft beer, and simple hefty meals. The owner, a jovial innkeeper, was happy to see Kuhn and offered his visitors a secluded comer booth. Having received two tall glasses of real draft beer from the host's secret supply, Kuhn began:

"My mother was an American who married a German diplomat before World War One. I was born in Washington, D.C. where my father was an Attache at the German embassy, but when America entered the Great War, we had to return to Germany.

"In July 1918, my father, an officer, was killed in the second battle of the Marne. After the war in 1919, my mother and I returned to the USA, living with relatives in Virginia. After finishing high school, I returned to Germany enrolling in the *Kunst Akademie*, studying art and graduated just before Hitler came to power. I really liked Germany. Later, I believed that Hitler and his Nazis would soon disappear again. Therefore, I remained a German citizen."

"However as luck would have it, some time ago I had a car accident and almost lost my leg. I survived the operation, but ended up with one leg somewhat shorter than the other, which is why I have not been drafted.

"What a lucky break!" said Peter.

"Just before this lousy war broke out, I had considered returning to the United States; because I had become extremely disillusioned with Hitler and his Nazis.

One of my social contacts at the American embassy was Richard Johnson, and eventually we became good friends. One day he asked me if I would be willing to stay in Germany and work indirectly for the State Department. He promised that after the war, I could return to the USA with honor, in view of my service. Being born in the USA would certainly allow me to easily become a US citizen. I was more than glad to serve my adopted country, the United States of America. Later when the US was drawn into the war, I was drafted directly into the OSS."

"My present assignment is to locate factories involved in the fabrication of these new rockets, harassing Great Britain, but my leads are meager. All I have found so far is a factory in Reinickendorf called Argot Motoren GmbH, which produces engines for the 'V-One' rocket. The Allies have been trying to bomb this place off the face of the earth, but still without success."

Mertens quickly but carefully considered, how much he should tell Kuhn on a 'need to know' basis.

"For your own security its better I'll tell you not too much about myself. However, our assignments are somewhat similar, but before getting you further involved, I'll need more information. It would be very helpful if you could provide me with the home address of Dr.Herbert Hassenkamp, a scientist working for the *Kaiser Wilhelm Institut*, who lives somewhere in Dahlem and is, as I understand, a devout Anti-Nazi.

Without a word, Kuhn walked to the bar counter and carried a thick Berlin telephone directory back to the booth. Quickly he found a Dr. Herbert Hassenkamp, living at *'Am Fischtal'* near the subway station Onkel Tom's *Hütte*. Mertens memorized the information, and told Kuhn he had obtained the scientist's name from Professor Rasche. He also revealed that Dr. Hassenkamp was working in the atomic development field. For the moment he didn't want Kuhn to know that Erika was Prof. Rasche's daughter.

"I must find a way to get to know this man without scaring him to death," Mertens said. But in a flash, he knew exactly how to solve his problem, and told Kuhn to do nothing at this point.

As agreed previously, one hour later Erika found them at the pub and all returned to the store-apartment. She and Peter collected their belongings and left the antique dealer with the understanding they would stay in close touch.

On the bus riding to Kopaneck's villa, Peter asked Erika if she knew Dr. Hassenkamp personally.

"Yes, I've met him and his family when the employees and their families gathered at the company cafeteria for Christmas parties and other social events."

Peter continued: "Tell me about what kind of person he is?"

Erika was surprised at Peter's interest in Hassenkamp.

"He is about forty-five years old, married to a nice lady from a Berlin army officer's family, and they have a daughter and a son, ten and twelve years old by now. Hassenkamp speaks with a pronounced Saxonian dialect, as he was born and raised in Dresden. I always found him a thoughtful man and very dedicated to his work."

For the rest of the trip Peter was quiet and thoughtful. On their arrival at the villa they met Wolfgang in the living room. Peter went upstairs to the two rooms he and Erika occupied. A modern tiled bathroom connected the quarters from either end. Wolfgang's master bedroom was further down the hall and included his own luxurious bathroom. From his suitcase Peter fetched his last bottle of Napoleon cognac and rejoined Erika and Wolfgang downstairs. The three relaxed and had a leisurely chat in front of the gas fire. Wolfgang told about his life, since the time Peter had seen him.

"As a result of my engineering education, the government placed me with a company in Reinickendorf, called the Argot GmbH, which produces engines. As you can imagine, our work is highly classified, and I shouldn't talk about it, but I can say, that what they've developed and achieved is mind-boggling. Fortunately, these projects are the main reasons why I haven't been drafted. However, I firmly believe the war is lost, and that Hitler's gang of brown-shirt opportunists will destroy Germany. My only interest in life at this moment is to survive this terror and the war unscathed, one way or another. It is really incredible to realize, how stupid and gullible the majority of the German people are. They love to shriek "hurrah" when the 'Teutonic Shrink Aryan' Goebbels declares total war against the whole world."

Peter knew he could trust Wolfgang, but also felt that his friend would not jeopardize his own safety or future by going actively against the Nazis. Wolfgang just proved Peter's contention; all he wanted was, to survive the war unharmed.

CHAPTER 13

The accommodations at Wolfgang's villa were an ideal setting for young lovers. The privacy of the interconnecting rooms helped a lot. It was late and time for all to retire to their bedrooms. Peter couldn't resist anymore his tempting desire to make complete love to Erika. In his pajamas he knocked at her door.

Somehow, she was half expecting him and answered: "Is it you, Peter?"

"Yes, I just didn't feel that I wanted to be alone tonight."

With these words he stepped into her room. In her light negligee, she immediately rose from her bed and walked toward him. Both embraced and became breathless after they felt each other's warm flesh through their thin night clothes. Passionately embracing, Peter lifted her in his arms and carried her to his room and bed. At first she was slow to respond because of her inexperience, but she couldn't resist Peter's advances. His good looks and clean masculine scent soon made her hungry for his body. Peter slowly undressed her and stripped himself. Soon bare flesh against bare flesh made them tremble with excitement and desire. Peter could not contain himself when he saw Erika's firm and proudly protruding bosom, her narrow waist, and superbly shaped long legs. Wavy, long blond hair curled over her naked shoulders, framing her classic face like a Greek goddess. Human passion took over; Peter had to make this Venus part of his life. To his thrilling surprise, he did not encounter resistance, and finally entered the gate to paradise. Peter felt overwhelmingly contented. He knew he had found the woman he had always longed for, someone he wanted to protect, forever.

Erika felt like she was swimming in a sea of passion, beloved by a man she saw as her hero, who would take care of her and who would one day be the father of her children.

After continued love-making and a blissfully relaxing rest, they talked about their future. Peter proposed marriage as soon as the war was over. Erika accepted joyously. If it hadn't been for the war, life couldn't have been more magnificent. They were lucky; there was no air raid during the night, spoiling their joyous love union.

The next morning they had the house to themselves, because Wolfgang had to commute to his work in Reinickendorf. There was plenty of food in the kitchen, so Erika didn't have to go out to shop; even though Kuhn had generously supplied them with food stamps and money for shopping in the neighborhood stores.

After Erika had prepared brunch in the large basement kitchen, she used the dumb- waiter to lift their meal up into the living room. After shaving, Peter got dressed. Erika was still wearing her bathrobe.

Peter asked as they sat at the table: "Erika, would you be willing to visit Dr. Hassenkamp tonight? I would go with you to see that you were safe. I want you to prepare the ground, so I can meet him in person as soon as possible. He has important information for us. Your father told me he is a devout Anti-Nazi and knows the *Gestapo* is looking for you both. I don't assume he would ever betray you."

"Peter, my meeting Dr. Hassenkamp would cause him some concern, and I hope he won't panic. But if you think it is absolutely necessary, yes, I'll do my best," she said.

"I understand, but you must try to persuade Hassenkamp to meet me alone at a convenient place of his choosing. Tell him I would like to talk to him about your father, but don't say anything about his successful escape into Switzerland."

Sometime after seven o'clock in the evening, Peter and Erika walked the few blocks to the subway station, Konstanzer Straße. Within fifteen minutes, they had reached *Onkel Toms Hütte* (Uncle Tom's Cabin). Apparently the story by the American author Harriet Beecher Stowe left enough of an impression on the Berliners, that they named a subway station and area in her honor; not even the Nazis changed that name.

The evening was cool, with a light drizzle permeating the air as they left the station and walked down *Waldhüter Pfad*, until they reached *Am Fischtal*. After checking the street sign for the proper range of house numbers, they turned left and continued walking. The doctor's house was about ten homes away, when Peter's keen eyes spotted, just around a slight bend in the street, a parked black Mercedes facing their direction. Stopping in his tracks before being seen, Peter stared at the vehicle, in which he could see two men sitting in the front seat, smoking cigarettes.

To be sure, he kept on walking with Erika, crossing to the other side of the street and bypassing Dr. Hassenkamp's home. Glancing at the Mercedes, Peter noticed the occupants of the car had disappeared; nobody was there. The only possible explanation was, that the two men in the car didn't want to be seen, and were crouching down in their seats, a move that made Peter very anxious and suspicious.

Peter and Erika kept on walking away from the car until it was completely out of sight, at which point Peter asked Erika to wait, while he'd return to investigate the mysterious black vehicle. Shielded by the side walk trees, he advanced carefully toward the Mercedes. This time approaching from the rear, making sure he would not be noticed. His suspicion proved right. Watching from behind a tree, he saw that two men were sitting in the car, one of them wore the usual shiny brown leather jacket. Both wore dark hats with wide downward sloping brims, the typical dress of a *Gestapo* agent. Obviously, they were scrutinizing the Hassenkamp home.

He retraced his steps back to Erika and both walked half a block south leading into a park, where they found a bench to sit and think the situation over.

"Erika, you can't visit the Hassenkamps; they're obviously being watched by the *Gestapo*, which makes it very dangerous to contact them. However, I have an idea which might work. We can see that the back of Hassenkamp's home borders this park. Although there is a fence all the way around the property, we could try to get to the back of their house from here. Under those circumstances, we both could visit the Hassenkamps, but first I have to carefully think this plan over before we make a move."

At this Peter scanned the park, making sure nobody was watching from that side. He walked along the bushes, assuring himself, that they were really alone. Finally Peter was prepared to act. They moved to get close to the rear of Hassenkamp's home, and approached the fence, which consisted of a cross-hedged wooden lattice, resting on a two-foot high concrete base.

There was no difficulty entering the backyard. Peter helped Erika over the fence, after which they quickly reached the door of a glassed-in porch, serving as a botanical display area.

Peter went ahead to knock at the door of the darkened porch. Fortunately, it bordered the living room where the Hassenkamps were sitting. A middle-aged man came to the glass door, peeked through the panes, and then opened the door slightly.

"Where did you come from?" he said. "Who are you? Why don't you approach my home from the front and in a civilized manner?"

While Erika stood out of sight in the shadows, Peter replied. "The front of your home is being watched and I want to talk to you in private."

"Nonsense! Who are you? What do you want?" the man asked, his suspicions highly aroused.

"I am a friend who wants to talk to you." At that moment Frau Hassenkamp entered the porch.

"Herbert, what's going on, who is this man standing out there?"

Now Peter waved Erika to approach the glass door. Immediately, the couple recognized Erika and opened the door. Before they entered Peter asked if the children were at home. Hassenkamp replied they were both in their rooms upstairs doing homework.

"Please keep them in their rooms. They should not know we are here," he suggested.

The couple understood his meaning and Frau Hassenkamp immediately went upstairs telling the children to stay put, as they had important business to discuss and didn't want to be disturbed. Both children obeyed, as they were used to strict discipline and obedience to their parents. Meanwhile, the visitors sat down in the cozy living room, where Peter introduced himself as an art dealer.

"Erika, for heaven's sake, where do you come from? How is your father? What are you doing in Berlin? And who is this handsome stranger with you?" Frau Hassenkamp seemed to have dozen of questions to ask.

Erika didn't know what to reply, because she thought she might say the wrong thing, so she looked to Peter to handle the situation, which he did by starting the conversation, speaking for them both.

'We wanted to meet you in a proper manner from the front door, however we discovered you are being watched by two *Gestapo* agents sitting in a black Mercedes across the street. That's the reason we approached you through the park from the rear."

"Erika's father, Dr. Rasche, is safe and sound in Switzerland, but before he left Germany he gave me your name, stating that you and your wife were ardent anti-Nazis. I am an art dealer having to take care of some important business in Berlin, but my normal home is in Strasbourg, where I had the good fortune to meet Erika. Dr. Hassenkamp, it is very important, that you and I sit down soon and have a quiet discussion. However, it would be very difficult to meet outside, because as I've said, you're being watched."

The scientist silently pondered Peter's words, while his wife left the room in order to sneak a look from her front windows, searching out the Mercedes. She returned, announcing that the Mercedes was indeed parked across the street, just as Peter had claimed. Dr. Hassenkamp shook his head in disgust.

"I assume, that as a result of our top secret military work, most of our scientists are being watched by the *Gestapo*, observing who we meet and to whom we speak. You were wise to have approached our home from the rear. But I don't know what purpose it would serve to have a meeting with you unless you can be more specific."

"I would like to discuss the status of your atomic research, because if Hitler should develop an atomic device, he could conquer, or even destroy the world."

Hassenkamp was shocked to hear this, but after a few moments of silence, he stood up from his chair and started to pace the room.

"You are asking me to commit treason," he said. "However, I agree with you, that Hitler would use such a weapon, ruthlessly, without mercy. He has already committed so many crimes, against the Jews, the Poles, against countless of our German intellectuals, many of them my close friends. I can no longer support this maniac. Mr. Mertens you are forcing me to make a momentous decision... My answer is... 'Yes', I will cooperate."

"Herr Doctor I appreciate your predicament, and I must admire your courage by making this decision, but please don't let me interrupt you."

"Here is what I would like you to consider," said Hassenkamp. "Next Saturday afternoon, when I take my weekly bicycle ride in the Grunewald, I would like to meet you alone at the Grunewald hunting castle, the place of our Hohenzollern Princes. We will meet at the main entrance. You'll need a bicycle to get there. Being an art dealer, you will provide me with a good cover story, in case the *Gestapo* should interrogate me. The story would be that you have offered me a painting by Max Slevogt, the famous German impressionist, and we are discussing the price and the deal."

"Yes, I'm familiar with Slevogt's works and I always liked his semi-impressionistic style," said Mertens.

"This painting, a portrait of Bruno Cassierer, the Jewish publisher and art dealer, was actually offered to me from another source for fifteen thousand Reichsmark," the scientist continued. "For your information, the painting is still available at the art dealer Kurt Heinrich in the Kurfürsten Straße. However, let's talk about this when we meet in the Grunewald. It would be better if you leave; please meet me at the castle at four in the afternoon. Auf Wiedersehen!"

Peter was delighted to hear Hassenkamp's brave decision, and he promised to appear at the suggested meeting place on Saturday.

In the meantime Erika told the doctor's wife about the reason her father had to flee to Switzerland from the *Gestapo*. Of course, the Hassenkamps knew all about this affair as it was known all over the Institute. Dr. Rasche and his family were publicly declared traitors to Führer and Fatherland and condemned to death in absentia.

It was about eight thirty in the evening when the howling of air raid sirens began their high-pitched up and down scale alarm cycles, a sound which was almost as frightening and eerie as the actual bomb attack. Peter and Erika said a hasty farewell and left the doctor's home the same way they had arrived, by climbing the fence and heading back into the park. They hurried in the direction of the above-ground subway station, Onkel Tom's Hütte.

On the way through the park they found a sheltered trench consisting of a simple ditch covered by concrete slabs with planted grass as top covering. The shelter could be entered from either side by an open sloped walk-in. It offered them and other people some degree of protection.

From the west, they heard the thundering sounds of the German anti-aircraft FLAK artillery with their deadly 88-millimeter guns. The British Royal Air Force was maintaining its routine nightly visit and Peter with Erika were watching the spectacle from the entrance of the trench.

Bright searchlights criss-crossed the sky seeking out targets; then, after a short period, a tiny object would appear high in the sky, glittering brilliantly like a star through the illumination of the cross beams. The brilliant spot tried to escape the FLAK's attention. Deep red flashes of anti aircraft shells exploded all around the single Pathfinder airplane, which eventually disappeared over the horizon, only to be followed by the frightening masses of attacking bomber formations.

In the distance appeared an array of numerous glaring red marker balls, grouped like an upside down Christmas tree, descending slowly to the ground. Now, a deep rumbling sound of bombs, striking aimlessly at the city of Berlin, was to be heard. Like a carpet, the bombs unfurled over predetermined sections, following a distinct pattern for the annihilation of the city, section by section.

Peter was deeply shocked and hypnotized by this brutal spectacle. It lasted about forty-five minutes, after which the sky was empty and silent again. A deep red glow of a fiery inferno appeared over the city, interrupted once in a while, by a sudden deep explosion and bright flaring flash, indicating a more volatile fuel source had been ignited.

The trip back to town was uneventful because the train ran underground before reaching the inner city. However, the walk from the station' to the villa was an eerie experience. Many buildings were on fire and a windstorm blew ashes in their faces, forcing them to squint to protect their eyes. The villa was unharmed. No more than a couple of incendiary bombs, each consisting of three-inch diameter hexagonal sticks, had landed in the front yard. Wolfgang had dowsed them with heaps of fine sand.

Wolfgang was glad to see his friends return unharmed, and all decided to spend the remainder of the evening in the living room. Peter asked Wolfgang if he could explain what kind of engines his company was manufacturing. But Wolfgang clammed up and changed the subject. Instead he talked about a party he had planned for Saturday night in his villa. He had invited some of his friends and a few women he had met at nightclubs. He was especially pleased to have hired a French accordion-playing singer to provide the background music.

The next morning Peter and Erika went to see Kuhn to tell him about the visit with Hassenkamp and the scheduled meeting next Saturday at the Grunewald castle. Kuhn was alarmed that the *Gestapo* was watching the scientist, and he warned to be extremely careful on Saturday.

"I believe I'd better tag along with you, as a back-up escort," he said. "Approach the scientist all by yourself, but it might be wise if I stay behind and cover both of you."

Peter thought the idea had merit and accepted the plan. Then he told Kuhn about the Max Slevogt painting and the fifteen thousand Reichsmark price tag. Kuhn knew about the deal, for it had recently been offered in an open auction and was now in possession of art dealer Kurt Heinrich, whose gallery was just around the comer at Kurfürsten Straße. Kuhn promised to talk to Heinrich, telling him he had a possible buyer. He would ask for a photo of the painting, thus providing a future alibi in case the *Gestapo* should ask questions.

Peter decided to search in the telephone book for SS-Standarten Führer Heiner Kempe at the '*Gestapo* Office for Jewish Affairs' in the Kurfürsten Straße. Asking Erika to keep dialing the connection, he eventually got hold of Kempe, who was as usual in an agitated state. After he learned Mertens was only a couple of blocks away from his office, he urged him to come over immediately.

The *Gestapo* building was nothing but a former five-story apartment structure, which had been converted to offices. Entering the hallway, Mertens reported to a desk sergeant wearing a black uniform with red swastika armband, who phoned Kempe and announced Mertens' arrival. After a couple minutes waiting, another SS-trooper appeared and escorted him to an open cage elevator, guiding him to a third-floor office. After knocking on the closed door, a shout from the inside: "Come in!" Kempe in his gray SS-uniform curtly ordered the trooper to leave, then got up from his desk and vigorously shook Mertens' hand. 'Welcome to Berlin! How about a cognac? Sit down!" After having produced tumblers and a bottle, they sat down in two leather armchairs.

"You have no idea how timely your arrival is." Kempe said. "Let me explain. Yesterday my boss Adolf Eichmann called me to his office, charging me with a very special assignment. I am supposed to organize a company of soldiers, including armored vehicles, to visit the *Alt Aussee* salt mines in *Die Ostmark* (the Nazi name for what was formerly Austria). Inside the salt mines are vast amounts of acquired valuable art treasures and gold bars, stored for safe keeping. I am ordered to select about eight cubic meters of the most valuable pieces, including all gold bars, and transport the whole cargo north to the Hamburg area. Since you are an art expert I could use your expertise to select the most valuable items. How would you like to join me on this exciting, profitable project? What do you say, Peter?"

Peter was amazed by this proposition and carefully thought the offer over replying: "It just happens that I will shortly conclude a business deal regarding the sale of a valuable painting. But before accepting your kind offer, tell me when this project is to take place and how long it will last."

"I'll need about a week to have a Platoon with armored vehicles organized and ready, following which we are supposed to deliver the cargo to a point on the Elbe River at the city of Wittenberg, where a motor barge will take on board the valuable freight. From there we will travel on the river through Hamburg to a place called Glückstadt, where new instructions will be given to us; therefore I believe we should be prepared to leave next week on Wednesday. We would drive in my new Mercedes, using Schaarführer Horst Stolz as our chauffeur. Remember, he's the one with the English officer's baton. In Hallstatt we would meet my company of soldiers. Unfortunately, I cannot tell you at this time how long this trip will last, but I'm authorized to pay you five thousand Reichsmark for your services. "

"Heiner, I accept your most interesting offer, and hope it will be a contribution to our war efforts. But what's the purpose of this project?"

"I don't know," Kempe replied. "But I am sure it is under orders from the Führer himself, who of course always knows what he is doing. I'll expect you to be at my office next Wednesday, let's say at seven in the morning, to start our trip." Kempe lifted his glass: 'To a successful conclusion of our assignment, prosit and Sieg Heil (cheers and Victory Hail)!"

After both men had finished their cognac, Kempe said. "And now I want you to meet Adolf Eichmann. He's most interested to make your acquaintance, I've already told him about you and our fantastic escape from Paris."

They walked a few doors down the hallway, passing another SS-guard station, and Kempe knocked at one of the closed doors. "Come in!" was the reply. After they had entered, Mertens saw a clean desk near the window where a man in a gray uniform was seated, wearing on his left arm the Nazi-eagle insignia. On the opposite wall hung a large picture of Adolf Hitler. Kempe started the conversation: *"Heil Hitler Herr Ober-Sturmbannführer*, may I introduce Peter Mertens, he is the man I told you about who engineered our escape from Paris."

Eichmann, who easily could have passed as an accountant or post office clerk, got up from his chair, walked over to Mertens, extending his hand in welcome.

"I am pleased to meet you in person," he said. "People like you are an asset to our great country. Have you ever considered joining our organization, the SS brotherhood?"

Before Mertens could reply, Kempe answered: "Herr Ober Sturmbannführer, Herr Mertens has just agreed to assist in our task assignment 'Golden Eagle', the code name devised for the high value freight transport from Die Ostmark. I'll have enough time and opportunity to persuade him to join our elite fraternity.

"Very well!" replied Eichmann; "I shall report our progress to the Führer.

Once you have reached Glückstadt, you'll receive further orders. We're happy to have you with us, Herr Mertens. Best wishes for a successful completion of the assignment, Heil Hitler!"

After they had left Eichmann's office, Kempe asked Mertens for a place he could contact him, in case of an emergency and Peter gave him the address of Wilhelm Kuhn, the art dealer. Then they said farewell to each other. Mertens confirmed he would be at Kempe's office on the following Wednesday morning.

On the way back to Wilhelm Kuhn's store, Mertens analyzed his impressions of the meeting. He noticed that Eichmann, who spoke with an Austrian dialect, looked anything but Aryan. He had dark curly hair, was of medium stature and easily could have been taken for Jewish. While at the university Mertens had previously observed, that many Gentiles who looked like, and could have been, Jews, were the most vicious anti-Semites. The only thing that bothered him was the desire by his new found "friends" to have him join the SS-Corps. He had to find a safe way to refuse this dangerous offer, for it could easily lead to the discovery of who he really was, and result in a disastrous ending for his important mission.

CHAPTER 14

In the middle of November, 1944, Mertens and Wilhelm Kuhn took two bicycles and traveled west along the *Kurfürstendamm* toward *Halensee.* The weather was cool and dry, the sun occasionally appearing through gray clouds.

They set off from Kuhn's place at two thirty Saturday afternoon; the distance to *Grunewald* was about ten kilometers. It was a most pleasant cycling experience. After crossing the Halensee Bridge over the *S-Bahn* or city train, they entered *Königsallee.* Pursuing its small-sized cobblestone pavement, they arrived at the Grunewald. Here they followed a marked sandy footpath pointing them toward the Jagdschloß or hunting castle.

The forest consisted mainly of tall fir trees with flaky, reddish bark. Once in awhile, a group of white birch offered a pleasant contrast. Due to the lack of underbrush the landscape more than anything resembled a park, especially since most of the trees were placed well apart. The ground was composed of light-yellow, loose sand, supporting a sparse grass covering, dotted with spots of open sandy pits. Napoleon called the Berlin scenery: "A large dusting powder pit, suitable for drying inked writings."

Occasionally, in the distance, the cyclists heard the sounds of barking dogs being walked by their masters. The woofing reverberated like an echo in a cathedral through the tall forest

They arrived at the castle about half an hour early. Just before reaching the entrance, Wilhelm separated himself from Mertens by riding toward the banks of the adjacent *Grunewald* Lake. Mertens always liked this mysterious castle, whose folklore told of a white-sheeted ghost haunting the place at midnight.

He walked inside the courtyard over its heavy round cobblestone covering, looking at the display of numerous deer antlers hanging from the building facades. Only a couple of other onlookers were in the area, and the rest of the forest seemed deserted.

Somewhat aimlessly, Mertens with his bicycle, walked toward the shore of the lake. The banks were bordered by a dense array of deciduous trees and underbrush. Every so often, the lake was accessible by a small sandy beach, where a wooden bench provided a view over the scenic forest shores and winding lake.

At four o'clock Mertens returned to the castle where Dr. Hassenkamp had just arrived. After a short greeting, they walked with their bicycles to a bench near the shore, where they took a seat overlooking the lake.

"What a beautiful restful area!" stated Mertens"

"I agree it is an unusual place and so near to Berlin." was the reply.

Carefully they spoke about the weather and other trivialities; finally Dr. Hassenkamp asked Mertens about his background.

"My father is German. He immigrated before the Great War to Chicago. My mother grew up in Quebec, Canada," he explained. "I attended the Berlin Technikum where I graduated as Diplom Ingenieur in spring of 1941. Afterwards I returned from Berlin to Chicago, joined the US Army and saw action in North Africa and Italy. Now I am here serving my country by utilizing my language capabilities and engineering knowledge. By the way, I was instrumental in getting Dr. Rasche across into Switzerland."

Dr. Hassenkamp was impressed by Peter's frankness. Intrigued, he started telling about his life.

"I always admired the United States of America for its wonderful freedom and boundless human energy, where anybody has a chance to make it to the top, regardless of his or her social background. Science and industry has made your country the richest and most powerful nation in the world. Only a lunatic like Hitler would challenge this might. If at all possible, I'm seriously considering moving my family to America once this war is over, and I hope through my cooperation with you to make it feasible."

Mertens was pleased to hear these words, because it reinforced his feeling that Dr. Hassenkamp was sincere, had an interest in Mertens' mission, and would offer his fullest support.

"I am impressed by your objectivity and willingness to prepare for your family's future." he said.

"It reminds me very much of my father's philosophy, which he told me in regard to Germany and America."

"I am proud to be of German heritage when it is based strictly on its culture and tradition of industry. However, politically I owe my allegiance to the United States of America and its Constitution."

"Dr. Hassenkamp, someday I shall do my best to help you and your family fulfill your dream."

After a moment of silence Mertens continued: "Doctor, what is the status of your atomic research program, after having lost your deuterium or heavy water, during the commando raids in Norway? Have you considered a substitute moderator? Or have you found other means of achieving fission?"

"Please understand," the doctor answered, "most of my prominent colleagues are utterly disgusted with Hitler and his brown gang. The thought of handing him an atomic bomb is unthinkable. So they are dragging their feet whenever possible. During a conference in February, 1942, Heisenberg was asked when a bomb could be made available. He replied that it would take at least a quarter of a century, before a workable device would be feasible. With this statement the Nazis lost interest and converted the effort strictly as a research project."

"Am I glad to hear that!" interrupted Mertens.

"Fortunately, Hitler is too stupid to grasp the theory of atomic science. He also shows certain contempt toward our group of scientists. I do not believe we'll develop an atomic bomb in Nazi-Germany," the doctor said.

"Let's assume you were interested in making a bomb; what kind of development would you recommend?" asked Mertens."

"Academically speaking though, we have considered the use of graphite as a moderator. We are very concerned how to control a successful chain reaction. We are afraid we might create a monster out of control, somewhat like *Goethe's 'Der Zauberlehrling'*, or Sorcerer's Apprentice. It would be like a train on a steep slope without brakes. Some of us thought the use of cadmium or niobium as a neutron-absorbing element could control the fission. However, as I said, we are only dabbling with these ideas."

"In other words, you are only theorizing and no practical efforts have been undertaken?" was Mertens' question.

"Experimental development work was placed in three locations. Werner Heisenberg, the founder of the modern 'Quantum Theory', heads one group. The lab is located in Leipzig. The second project takes place at the KWI (Kaiser

Wilhelm Institute) in Berlin which also is headed by Heisenberg, von Weizäcker and Karl Wirtz. The third group, under Diebener, is located at the Army-Arms-Department in Berlin's western suburb of Gatow. "

"Wouldn't it be possible that one of these extensive efforts could show some unexpected success?" was Mertens' question.

"I am sure nothing will come of it before the war is lost. Another group of our scientists believes the uranium atomic weight of U-238 might not be the entire story; but we haven't found any means to analyze uranium or its characteristics. There is still much to be learned."

Mertens was fascinated with the Doctor's comments and anxious to find a way to convey this important information to Richard Johnson in Zürich. Maybe Wilhelm Kuhn has some method of transmitting vital information to Switzerland.

While Mertens and Hassenkamp were having their meeting, Kuhn walked into the forest, trying to find a spot from where he could observe the two interlocutors. Pushing his bicycle uphill, away from the southern shore of the lake, he thought he saw a glistening mechanical object ahead of him.

Carefully, he continued to climb the slope until he found a spot among a clump of birch trees, where he could hide his bicycle. Approaching without being seen, he came across a black Mercedes parked among some trees. The car was hidden by surrounding bushes, and was angled downwards toward the lake. Skirting the vehicle and shielded by the bushes, he finally came up on it from behind, always relying on the fact that the rear view mirror of the car would be angled upward.

Making sure he would not be seen, he studied the two men sitting in the front seat behind their closed windows. They matched Mertens' description of the *Gestapo* agents who previously watched Dr. Hassenkamp's home.

Wilhelm Kuhn was worried that the scientist was still under scrutiny; it meant the *Gestapo* would now begin to investigate Mertens, possibly arresting him for questioning. To his dismay, he noticed one of the thugs scanning Mertens and the Doctor on the bench, via a pair of binoculars. Now something had to be done immediately.

Surveying the surrounding area Wilhelm saw a pile of dried up pine tree branches. Suddenly, Wilhelm had a great idea! Still unseen and exercising extreme caution, he carried some of these small branches to the back of the Mercedes. After being satisfied with the amount gathered, he returned to the birch tree area, where he picked up several pieces of bone-dry, papery birch bark. When he had collected enough, he returned to the automobile and meticulously pushed the branches underneath the rear of the car, between the two back wheels and exactly under the gas tank. Then he placed the birch bark, twigs and leaves beneath the pile of small pine branches so they almost touched the bottom of the gas tank. Taking a match from his pocket he lit the bark and immediately started a fire. Wilhelm retreated as fast as he could, scrambling to his bicycle and peddling furiously down to the shoreline footpath. He halted about a hundred yards to the left of the bench where Mertens was talking to the Doctor. Just when he was getting off his bicycle, a huge explosion roared through the forest lifting the rear of the Mercedes high into the air, flipping the car on its back and engulfing the Mercedes and everything in the surrounding area in a sea of flames. Convinced that the two agents could no longer be alive, or at least completely incapacitated, Wilhelm rode his bike over to Mertens shouting: "Let's get the hell out of here!"

Startled, the two men grabbed their bikes and followed Wilhelm out of the *Grunewald* in the direction of Konigsallee and back to town. Exhausted, they reached Wilhelm's gallery. Finally, they rested in the living room, where Wilhelm offered a glass of cognac, helping to calm everyone's nerves.

Still being cautious, Kuhn suggested taking a stroll across to *Wittenberg Platz*, where they found a comfortable bench. Now he had time to tell how he had managed to put these *Gestapo* agents out of action, and before they could discover Mertens' true identity and mission.

Mertens painstakingly thought this situation over and finally voiced his opinion. "Wilhelm, you did the right thing; because the *Gestapo* is still unaware of my existence. The main thing I'm now concerned about is the safety of Dr. Hassenkamp and his family. I imagine he could be tortured by the Nazis to discover if there was a connection between him and the explosion. We must come up with a solid cover story for the doctor, freeing him from any suspicion."

Turning to Dr. Hassenkamp, Mertens asked: "Do you regularly visit the Grunewald Lake on Saturdays?"

"Yes always, I come to exercise and to stay physically fit."

"Undoubtedly the *Gestapo* has seen you before at that location; that means we have the basis for a good cover story," Mertens said.

"Doctor, you told me about the oil painting by Max Slevogt," he continued. "I believe this could be the basis for our story, which could be like this: Wilhelm Kuhn met you at the Grunewald Castle to show photographs of the painting and negotiate a price. It was he you dealt with, while sitting on the bench at the lake. As for myself, I never existed.

"Let's hope we can fool the *Gestapo* with this story," Kuhn interrupted.

"We can't ignore the fact that Dr. Hassenkamp was seen by someone," Mertens continued. "I saw two other visitors at the castle who would probably be found and interrogated to report their observations. In the meantime, Wilhelm as a precautionary step has already contacted the art dealer, Heinrich, in the Kurfürsten Straße and obtained a photo of the portrait painting. Wilhelm, please give it to the doctor when we return to your place."

Later, as agreed, Kuhn gave Hassenkamp a page-size black and white photo of the art dealer Bruno Cassierer's portrait. Then he said: "By the way, the price suggested for the painting is now twenty five thousand *Reichsmark*."

"Doctor, I believe this will be a good alibi," Mertens said.

"You can use this cover story to prove you really negotiated for the portrait with a legitimate dealer. We'll keep an eye on your family and do anything to protect you, when and wherever possible. By the way, I also have intimate contact with the *Gestapo* and will know what's going on. This is just part of my job. Whatever happens, you must trust us completely. In an emergency, you can always ask for help from us right here, either in person or by phone, but be careful when calling, as the line might be tapped."

Still apprehensive, shaken and not fully convinced, Dr. Hassenkamp took his bicycle and left for home.

Arriving back at the villa, Mertens noticed that quite some activity had taken place in his absence. He remembered his friend Wolfgang Kopaneck had planned a party for that evening in the villa.

Wolfgang had made all the preparations, selecting the large dining room on the ground floor for the upcoming event. The oak parquet floor and a couple of crystal chandeliers overhead, added to the room's overall elegance. Three French doors led into the brick wall-enclosed courtyard, which was mostly planted with roses, supported by white trellises.

The first guests arrived at eight in the evening. The party eventually included eleven in all, namely three other male co-workers from Wolfgang's factory, with four chorus girls from a nightclub and the *Metropol* Theater. Also present was the Frenchman Jean-Paul, and of course Peter, Erika, and host Wolfgang.

The accordion-player was excellent, providing a mixture of French traditional *Musette* music and well-known German hit tunes. When he took a break, Wolfgang changed the musical entertainment with phonograph records. They were mostly American or English bands. On the electronic seventy eight-rpm player, he relayed all through the loudspeaker system.

As the evening progressed the guests became more and more cheerful and noisy. Everyone was either dancing or indulging in the first stages of love-making. The room soon filled with a blue haze of smoke. Wolfgang supplied the guests with a generous amount of 'Navy Cut North State' British cigarettes, which left a sweet, perfumed smell in the atmosphere.

Eventually, a couple of the girls started peeling off their clothes, while dancing alone. One beautiful brunette wore nothing but a grass skirt as she imitated a Hawaiian Hula dance. Her young body was perfectly shaped, and her breasts protruded challengingly in the air, vibrating to the rhythm of the music. Once in awhile one of the male guests would disappear with one of the girls, retiring to the empty bedrooms upstairs. At this point, Peter and Erika left this most lively party, retreating to their own quarters.

At two o'clock in the morning, air raid sirens announced another British attack. Fortunately, the targeted area was far north of the city and nobody seemed to care, as long as they didn't disturb the party.

Early on Sunday morning the telephone rang. Wilhelm Kuhn urgently wanted to speak with Peter who, still drowsy, went to the telephone.

"Peter I just received a call from *SS-Standarten Führer* Kempe who wants you to contact him immediately. Here is his telephone number."

Peter acknowledged the request and after finishing the conversation, dialed Kempe's number. On the second ring he answered: "Peter I want you to meet me right away at my office in the *Kurfürsten Straße*. Something terrible has happened, and I need your help."

Within half an hour, and after checking with the desk Sergeant, Peter was in Kempe's office

"Peter, sit down and listen!" Kempe said. "Yesterday afternoon two of our agents were murdered in the *Grunewald*, their Mercedes set on fire, and the gas tank exploded before they could get out. Normally my office would not be involved in this type of investigation, but there is a twist to this affair. It concerns the purchase of a supposedly valuable painting by some degenerate German artist. I believe his name is something like Max Voigt. We have arrested one of our top scientists, a Dr. Hassenkamp; because he was the person the two agents were assigned to watch. We have been trying to find a connection between this man and the assassination, but he seems to have an alibi, which is that he met an art dealer in the *Grunewald*.

"This art dealer, by the way is your business partner Kuhn, who gave Hassenkamp the photograph of the painting for consideration. He mentioned the purchase price of a ridiculous twenty-five thousand *Reichsmark*. Dr. Hassenkamp is presently held for interrogation at our headquarters in the *Prinz Albrecht Straße*, and I want you to come with me right away to assist as an art expert during the inquest."

"I also intend to interrogate your partner Kuhn. Maybe he can shed some more light on the situation."

Peter was at first speechless. Pulling himself together and recovering from the shock, he said: "Heiner, I am pleased that I can be of service to you! I knew Kuhn had a date with someone, who wanted to buy this painting. Kuhn acted strictly as an in-between agent. But in my wildest imagination, I can't see any connection with the ghastly deed you have described."

Both men went downstairs to the Mercedes, and Kempe drove them to Prinz Albrecht Straße. The *Gestapo* headquarters was a frightening place, even on a Sunday morning. Once inside, many hallways with separating steel bars and gate control points had to be passed. Finally, they reached the office of the interrogator, a Herr Roland Stoffeld. Stoffeld was a stocky, tough-looking character, with intelligent light blue eyes, wearing a shiny brown leather jacket. He used to be a civil service *Kriminal Inspektor* with the city police department, but was drafted into the *Gestapo*. He was a no-nonsense professional.

Kempe introduced Mertens as an art expert, and after shaking hands Stoffeld explained the details of the case, saying that they did not have proof of Hassenkamp's wrong-doing. However, they thought no stone should be left unturned, especially since the scientist was under surveillance by the two agents, just prior to the assassination. Kempe asked Stoffeld: "Do you have any anti-state evidence against Hassenkamp, like defaming our *Führer* or other defeatist remarks? Has he been associating with Jews, or does he have any anti-Third Reich friends?"

Stoffeld was prepared to answer this question: "We have no concrete evidence, but he is not an active supporter of the Third Reich and doesn't seem to be interested in politics. However, the fact that he wants to purchase a portrait of this Jewish swine Bruno Cassierer doesn't speak well for him at all."

"Let's go and talk to him. Maybe we can uncover some hidden skeleton," Kempe said.

Stoffeld agreed and together they walked through another gate point where one of the guards led them to a special room. Mertens was extremely apprehensive that Hassenkamp would reveal his true identity, under the wrong impression that he, Mertens, had betrayed him.

The guard unlocked the door and the three men entered a semi-darkened room with a turned-off spotlight standing in the center. In one corner stood a wooden slatted rig shaped like a barrel in horizontal position, to which a number of leather strips were attached. A vicious-looking leather strap whip rested on top of the rig.

Stoffeld switched the spotlight on, directing the light into Hassenkamp's face. The Doctor sat slumped on a hard wooden chair, his glazed eyes revealing hopeless fear.

Stoffeld started the interrogation: "We have found evidence that you are connected with a group of Communist bastards who are plotting to harm our country!"

Hassenkamp's reaction was instantaneous. He loathed Communism as much as the Nazis:

"This accusation is insane. I hate Communism and I never have nor will I ever associate with these primitive *Untermenschen*, (sub humans)."

"You must have heard the explosion in the forest," Kempe asked. "Why didn't you go and investigate the cause?"

"You know I was talking with Kuhn, the art dealer. We both wondered what had happened, but didn't want to get involved in whatever caused this explosion, so we took our bicycles and left for the city."

"What possessed you to negotiate for this degenerate piece of Jewish art?" Kempe asked.

For a moment Hassenkamp was at a loss for words, but then answered: "I don't think Max Slevogt is a degenerate at all, nor is he Jewish. He is highly-acclaimed in professional German art circles through his illustrations of various books, including a number of magnificent paintings. I never gave it a thought, that Bruno Cassierer was a Jew. I only wanted to buy a valuable painting and thought it would be a good financial investment."

At this point of the interrogation, Kempe turned to Mertens and asked him to check if these statements had any merit. Mertens decided to step forward into the lighted area, so he could face Hassenkamp directly; while blocking the view of the onlookers.

Hassenkamp almost choked when he saw Mertens, who immediately twinkled his left eye, raised both his open hands at chest height and signaled a short, but limited wave motion. Then he jerked his head in a hushing expression that said: "Don't give me away!" Hassenkamp seemed to understand Peter's message and calmed down somewhat. He suddenly remembered Mertens' hint at the bench on Wittenberg Platz that Peter was intimately in contact with the *Gestapo*.

Knowing the Prussian mentality, Mertens shouted at Hassenkamp. "You ignorant misfit, don't you know that even admiring this painting involves you with a Jew? Have you already bought this piece of trash?"

"No, but in light of your explanations, I certainly am no longer interested. "

Having heard enough, Stoffeld switched off the light and the three men left the scientist alone in the room to return to Stoffeld's office.

"We don't really have anything we can pin on this man," Stoffeld said.

I understand too, that he is an important member of the *Kaiser Wilhelm Institut*. We better not hold him any longer, as he must return to his classified work. But we shall continue to keep an eye on him." Then he turned to Mertens and asked: "Tell me about this Max Slevogt. Who was he, and is he still alive?"

Mertens, remembered his German impressionists and coincidentally had read a book about Slevogt he found at Wolfgang Kopaneck's villa in its extensive library. Therefore he said: "Slevogt was born in 1868 and died in 1932. He illustrated a book of Mozart's *'Magic Flute'*. He also painted many landscapes and portraits, including 'Bruno Cassierer', who was one of the most prominent art publishers and dealers, not only in Germany, but also throughout Europe.

Fortunately, through Adolf Hitler. Germany has now been cleansed from his

Jewish influence."

Stoffeld thanked Kempe and Mertens for their visit and advised them that as soon as they had gone, he would send Dr. Hassenkamp home to his family.

With an enthusiastic mutual Heil Hitler salute the men separated and Kempe drove Mertens back to his office. He reminded him of their Wednesday date to leave for *Die Ostmark*, and then the two men parted company.

Back at the villa, Mertens decided to leave Dr. Hassenkamp alone; realizing any contact with him could prove fatal for both of them. He worried only about possible suspicions against his partner Wilhelm Kuhn. But Peter knew Wilhelm would be able to take care of himself.

The next day he visited him and both decided to walk through Berlin's *Tiergarten* Park, where they knew, they would have privacy. Now it was even more likely the *Gestapo* would keep Wilhelm under scrutiny and his home would be bugged. Caution had become essential. Mertens gave Wilhelm the details of his atomic science conversation with Hassenkamp, and asked him if he could get this vital information to Richard Johnson in Zürich. For this he gave him a note he had prepared, to be sure the information would be relayed correctly.

"No problem," Wilhelm replied. I have a friend in one of the neutral embassies. He'll relay these data as soon as possible to Johnson via diplomatic channels."

Mertens reminded Wilhelm they had both promised to help Hassenkamp after the war was over, urging him in the meantime, to take good care of the doctor's family.

CHAPTER 15

The weather in Berlin was dreary and gray during November 1944, with intermittent rain showers not improving matters. In the two-day period, while Mertens had been involved with the Grunewald affair, Erika had followed his advice and enrolled as a voluntary nurse in the Westend Military Hospital, located right across from the Heerstraße S-Bahn station.

The hospital, with its huge Red Cross painted on top of the roof, used to be a school building, but had been converted to accommodate wounded soldiers arriving from the multitude of combat areas. Peter thought if Erika served as a voluntary nurse, wearing the nurse's uniform, she would avoid suspicion or questioning from neighbors and other snooping citizens.

Back at the villa and alone in their quarters, Erika urged Peter to listen. She was very concerned for his safety.

"Tomorrow morning you'll leave with these two dangerous men on a lengthy trip," she said. "This cruel fellow Stolz will be with you all the time and he is the one you have to watch out for. All he needs to remember is the day in 1941 when he came to your apartment with the *Gestapo* and hauled away this poor Jewish family."

"I don't think I'm in too much danger," said Peter. "Stolz saw me in my room while it was still fairly dark outside. I had my window open because of the blackout; therefore I wasn't allowed to switch on my room lights. Stolz used his flashlight to glance at my American passport and handed it right back to me. The only reason I remember him is his silly baton and pronounced peasant dialect. Be assured, I'll watch and if the slightest suspicion should crop up, I'll be prepared to defend myself."

Erika was extremely concerned. She clearly analyzed her situation; the more she shared her life with Peter the more she realized what an unusual man he was. The few German men she had previously known treated her very differently. They always acted more like they were a gift to every woman they met. Rarely, did they show the same courtesies she so often noted with Peter.

Early on Wednesday, after an emotional farewell, including a promise to return and take care of her, Peter took the bus to Wittenberg Platz from where he walked to the SS-office in the *Kurfürsten Straße*. Kempe was waiting for him in the entrance hall. After the routine Heil Hitler salute, they walked to the parked Mercedes on the street where *Schaarführer* Horst Stolz, the body guard and chauffeur, was waiting

Shortly after seven in the morning they started their trip south in the direction of Munich. It was an uneventful drive and gave Peter and Kempe plenty of time to talk with each other; while Stolz silently listened and drove. Peter learned that Strasbourg had been evacuated, and the Americans now occupied the entire Alsatian area west of the Rhine River. Peter thought he should have left Erika at the apartment of Mme. Abel; however, he was glad he had brought her to Berlin; otherwise, they might never have become such wonderful lovers.

Kempe was in a talkative mood, hinting that soon an important meeting would take place at the *Führer's* chancellery in Berlin, to which he was invited.

"This meeting will be the turning point for Germany. Hitler, in his genius, will surely reveal how our secret weapons will be deployed, and the war decisively turned in Germany's favor. Once victory has been achieved, Germany will create a new Europe with Berlin as its capital. The *Führer,* with the assistance of Albert Speer, has already designed a beautiful Capitol Building. All German-speaking, and German-related countries, will become part of Groß Deutschland (Great Germany), which includes Switzerland, the Netherlands, and all of Scandinavia. We are going to create a race of outstanding people and only healthy and suitable citizens will be permitted to have children. Sick and deformed creatures will be done away with. Of course, Jews will no longer be a problem, because they just won't exist anymore."

Keep on dreaming, Mertens thought. Soon you and your sadistic driver and the whole rotten Nazi-Neanderthal establishment will get what they deserved long ago.

Toward the evening they arrived in Munich, where they stayed overnight in the hotel Vier Jahreszeiten (Four Seasons). The city was in a bad state; because these day and night raids had reduced many apartment buildings to nothing but empty, burned-out shells.

Kempe chose to have dinner at the Hofbräuhaus in honor of his *Führer.* He wanted Mertens to see this historic place, where Hitler had begun his career, by speaking to the citizens of Munich. The Hofbräuhaus was a huge hall with many plain wooden tables, benches and chairs. Most of the guests were army personnel drinking their IG-Farben Ersatz beer, a concoction of chemicals simulating the color and taste of the brew, but without alcohol content. As far as Mertens was concerned, the place looked drab and uninspiring.

Next morning at nine o'clock, after a meager breakfast consisting of two rolls with marmalade and a pot of barley coffee, they left Munich heading for Salzburg. The temperature was quite chilly, hovering around freezing, while the gray sky indicated the possibility of snow. Near the town of Freilassing, the first snow-covered Alps appeared. Shortly afterwards, they crossed the old Austrian border, and arrived in the fairy-tale-like city of Salzburg, dominated by its formidable mountain top castle.

Passing through the cobblestone-paved city center, they reached a country road leading toward Mond (or Moon Lake), which is surrounded by a magnificent profile of lower Alps, skirting the lake's west shore. While the beauty of the scenery awed Mertens, Kempe was completely oblivious to the surroundings and couldn't wait to return to Berlin and his beer-drinking Nazi comrades.

The trip continued through a small valley until they reached Wolfgangsee, the next lake, near the town of St. Gilgen, at an altitude of some two thousand feet. Passing the southerly shore of Wolfgangsee, the car drove upwards in a steady climb. The temperature fell below freezing, the Mercedes rolling and bumping over the slippery snow.

"What ever happened to those French hostages you wanted to round up?" Mertens asked suddenly in a grave voice.

"Good old Stolz took care of that matter, right Horst?"

"Yes, we took good care of them; those damn Frenchies will think twice before they murder any of our German comrades again; however, we still have some unfinished business to attend to. They even tried to kill me, but I am too tough, and survived their cowardly attempt to assassinate me. Unfortunately, my comrade, Hochstetter, was not so lucky. But I know who these bastards are, and as soon as this war has been won, I shall return to Strasbourg, hunt them down, and kill them all."

As they continued their journey upward, they passed through areas, where both roadsides were dominated by imposing snow-covered high alpine peaks.

Shortly after midday, they reached the famous spa, Bad Ischl. It was here where the composer Franz Lehar wrote many of his famous operettas. The town was still charming and untouched by the war. It surrounded a well-maintained park, centered by its stately Kurhaus (spa building).

On the main street they parked and entered the Cafe Zauner for a lunch break, the snow continuing to fall heavily. The cafe was filled with older people, predominantly women, who, in most cases, had escaped from the big cities and their constant air raids.

After lunch, driving less than half an hour, carefully negotiating the slippery snow-covered road, they reached Hallstätter Lake. This was a most impressive body of water; especially unusual was the southwest end of the lake. The almost vertical, wall-like rock formation provided a backdrop for the quaint village of Hallstatt. After passing through a dark mountain tunnel, they reached their destination, where Kempe had an appointment with a company of SS-soldiers, including their armored vehicles and trucks.

Together with two armored vehicles and five trucks; outside the village on the lake's south shore, a number of bivouacs were erected. After Stolz parked the Mercedes among the vehicles, he located the *Schaarführer* in charge of the SS-company, who immediately walked over to Kempe and reported: "Heil Hitler Herr *Standarten-Führer*! Obediently I want to report, that the Golden Eagle Company has arrived and is ready to proceed on your orders!"

Kempe acknowledged the report and advised the NCO to wait for further instructions. He then turned to Stolz and told him to locate the town's cemetery superintendent Anton Gruber, who also served as gravedigger. "Order him to immediately come to the village hotel," he said.

The hotel was located in the cobblestone village square. Two rooms were reserved for Mertens and Kempe. It was a charming, quaint building with white-washed exterior walls and a colorful painting of Saint Christopher, carrying the Jesus child on his shoulders across a river. After Mertens and Kempe had checked in, they met again in the hotel restaurant, where they waited for the cemetery official and Stolz.

Anton Gruber was about fifty, stocky, clad in a pair of worn corduroy pants and heavy mountain boots, which were densely studded with thick steel nails. Gruber wore a Lodenfrei greenish jacket, with deer antler bone buttons, dark green collars and sleeves. His trembling hands clenched a light gray, short brimmed felt hat, adorned by a chamois brush ornament. His face was weather-worn, tanned, and creased. His gray eyes focused uneasily at the floor, and occasionally toward the men who had ordered him to appear.

After the usual Heil Hitler salutes, Kempe said, "I understand you have three coffins stored on your premises, for which you have written orders to release them to my control. Is this correct?"

"*Jawohl Herr Standarten Führer*!"

"Tomorrow morning at seven o'clock I want you to come to this hotel and lead us to the caskets, do you understand?"

"*Jawohl Herr Standarten Führer*! I shall be here as requested!"

After Gruber left, Kempe instructed Stolz to order thirty of the strongest SS-soldiers to appear at seven in the morning, outside the hotel, together with three canvas-covered trucks, and then dismissed Stolz to carry out his orders. Mertens was startled by these developments, and couldn't understand what Kempe had in mind with three coffins, but decided to keep quiet while waiting for the next day's events.

Next day, after they had consumed their early breakfast, Stolz, the SS-soldiers, and trucks arrived. Gruber came to lead the group through narrow streets and uphill lanes, climbing hundreds of steps, to the village cemetery. The climbing path over the snow-covered ground was steep. Eventually, the mountain's vertical rock formation indicated the end of the climb; they could go no further.

Reaching the graveyard and its adjacent church, Mertens was surprised by the small-size cemetery, containing only a few graves. The view from this spot, however, was breathtaking. Below lay the village of Hallstatt, with its medieval and tightly-arranged buildings, appearing like a display of Christmas toys. The formidable lake and its alpine background completed the picture of idyllic magnificence.

With their trucks parked below in the Village Square, thirty SS-soldiers followed the lead group. Now they assembled in front of the cemetery entrance, awaiting Kempe's latest orders. Guided by the gravedigger, Mertens and Kempe reached the charnel house, where the bones and skulls of the dead were stored.

In his Austrian Mountain dialect, which sounded like a twisted nasal grumbling, the guide explained the purpose of the storage system. Due to the very small size of the cemetery and the lack of tillable ground, the villagers, over centuries, had devised a system; whereby buried bodies remain no longer than twelve to fifteen years in the ground, depending upon the demand for new space in the cemetery. When a new funeral is announced, the gravedigger will open one of the oldest graves, digging until he reaches the rotten coffin. He will then remove the remaining lumber to be used as firewood. The skull and bones he'll wash and dry. Then he'll stack them inside the charnel house. He will mark the skulls with the names and dates of the deceased. Upon special requests from relatives, he'll paint artistically bright colors, flowers and ornaments on the forehead of the skulls.

With apparent pride, the gravedigger led the visitors inside the dome-shaped structure of cemented fieldstone. It was called the charnel house. At its cave's far end was the main storage area, consisting of an array of racks, on which hundreds of skulls and bones were neatly stacked.

Mertens was amazed, but before he could say a word, Kempe briskly ordered the gravedigger to lead him to the three coffins. Without a reply the cemetery man left the charnel house, and followed by his visitors, led them to an isolated fieldstone dwelling, used as an equipment shed. The three men entered.

In the semi darkness of the place, three coffins stood on the stone floor. Out of piety, the gravedigger had placed two lit candles on top of each casket. The shadows of flickering light, along with the shivery cold, damp silence of the place, cast an eerie atmosphere over the men. Respectfully, the cemetery official looked at Kempe and said: "*Herr Standarten Führer*, I know these coffins were brought to this place by another group of SS-men, but I noticed they were awfully heavy, what kind of bodies are inside the caskets?"

Kempe became angry and shouted at the gravedigger: "Mind your own damn business and don't ask questions!"

Then he walked outside, and called Stolz and his assembled soldiers to enter the storage shed.

With brutish indifference, the soldiers flung the candles from the casket lids, and ten men per coffin started to lift. The SS-men strained under their load, and finally managed to move the heavy burden outside into the open, walking carefully down toward the village. Twice they had to stop, putting their loads on the ground, catching their breath and regaining new strength. Eventually, reaching the cobblestone square, they succeeded loading the coffins in two trucks. The group drove off from the village in the direction of their bivouacs. Kempe and Mertens returned to their hotel, picked up their belongings, and followed the trucks in their Mercedes.

Under orders from Kempe, the caskets were off-loaded and stored inside one of the tents. At the rear, one of the armored vehicles was backed close to the opening, preventing outside viewing of the upcoming activities. Kempe took Mertens inside the tent, where three selected soldiers were busy opening the coffins. After the lids had been pried off, a white sheet was removed covering the contents, and to Mertens surprise he saw each coffin filled with shiny gold bullion bars, stamped with markings in Russian. He counted at least twenty ingots in each casket, and estimated the weight of each loaded coffin at about 400 kilograms or 880 pounds; no wonder the SS-men were straining under the load!

After half of the ingots, from all three caskets, had been transferred inside the armored vehicle, its rear gate was closed and the other armored unit was brought into the same position, to be filled again with the remaining precious cargo. In the meantime, the soldiers built a bonfire, where the empty coffins were consumed by brightly crackling flames.

In accordance with Kempe's prearranged plans, the company decamped and the convoy of trucks, guarded by armored vehicles, drove off toward the village of Alt Aussee. Bypassing a valley between the Dachstein formation and the High Sarstein Mountain, they drove until they reached the nearby village of Bad Aussee, a quaint village located inside a deep valley.

After another fifteen minutes, they arrived at the lovely resort town of Alt Aussee, located directly on a small, charming lake of the same name. Near the shore they found an ideal place for the soldiers to pitch their tents, while Mertens and Kempe used the rooms in the adjacent hotel.

Right after lunch, Kempe called Stolz to put snow chains on the Mercedes and be prepared to drive them to the nearby salt mines. At the far end of the village, a narrow snow-covered gravel road led uphill into a magnificent, fairy-tale forest of tall spruce trees, all covered by a white blanket of snow. The Mercedes had a hard time moving through the deep snow, but the chains coped successfully and soon, through the trees to their left, they saw a whitewashed stone building .. It was named the Steinberg Haus, and served as the entrance to the salt mines.

Stolz located the mine Superintendent who had been informed of their arrival, and the visitors were guided to the tunnel entrance situated at the end of the buildings, from where the salt mines could be reached. A couple of policemen in green uniforms guarded the mine opening. Upon their arrival, everyone was provided with a miner's helmet and its attached battery-operated lamp.

On one of the walls inside the building hung a map detailing the entire salt mine layout. Stopping at this point, the Superintendent explained the location of the treasure storage areas deep inside the mountain caves. The two main tunnels leading down were provided with a narrow gauge, electrically-driven train system, bypassing a multitude of smaller side tunnels, which led to a large number of individual salt caverns. However, their objective was to reach two main caverns, one named Kaiser Josef Werk, the other Zirnfeld Werk.

Under direction of the Superintendent, they seated themselves on a vehicle behind the electric locomotive. He drove the train downward into the mountain. After only eighty yards, the track divided in two directions, with the locomotive taking the right branch. The trip lasted about another three hundred yards before arriving at a looped turn, where they left the train.

Their guide lit the flames of a hand-held torch, and they walked another fifty yards through a tunnel that eventually led to the huge Kaiser Josef cavern.

The size of the place and the sight of the hoard of art treasure staggered Mertens. Many were crated and stacked in rows over the salt stone floor. The Superintendent explained that the storage was initiated by Adolf Hitler in late fall of 1942 to safeguard all these treasures from bomb attacks. Also, the *Führer* wanted to use the collection as a basis for a gigantic art museum to be erected in the city of Linz, honoring the place he was raised as a child.

There were 6,577 masterpiece paintings in storage, together with hundreds of sculptures, classic furniture, coin collections, jewels, and a large library of valuable books. Exceptional treasures like Breugel's "Peasant Wedding," Albrecht Durer's portrait of "Emperor Maximilian," a self-portrait of Rembrandt, the "Altar of Ghent," and innumerable other precious items had been pilfered, collected, and stored there. Much of the loot had been stolen through the so-called, Alfred Rosenberg Action, removing art treasures from conquered countries in the west, especially France.

Through the dim light of the burning torch, Mertens inspected many of the art pieces, his thoughts racing as he started to plan for the removal of the most valuable items. Finally, the Superintendent guided them to the far end of the cavern, where they had to climb at least twenty more steps, leading to another tunnel. After a walk of about one hundred yards, they reached the Zirnfeld cavern, where additional priceless works of art and other treasures were stored.

Kempe constantly asked Mertens for his opinion, which of all these valuables should be considered for the removal and transport. Knowing that transportation could easily damage these delicate treasures, Mertens concentrated mainly on small bronze statues, coin collections and smaller paintings. Having made decisions and selections, it was agreed to prepare special labels identifying the items selected for transport.

Returning to the Steinberg Haus, they instructed the Superintendent, through the use of the railroad system, to aid in the removal of the chosen items the next day. Kempe would supply his soldiers' manpower. At this point the Superintendent asked Kempe for his help and advice, regarding a very delicate matter involving Eigruber, the region's *Gauleiter* or party leader. Kempe and Mertens followed the man into his office. After they were seated, the Superintendent related the following story:

"*Gauleiter* Eigruber has placed eight one-thousand-pound American bombs inside the art storage areas, in order to eventually blow up the entire art collection and the salt mines. His reasoning is that in the case of Allied Forces arriving, all treasures and valuables must be destroyed before international Jewry could take its possession. *Herr Standarten Führer* I implore you, talk to the *Gauleiter* and change his mind. The preservation of the salt mines alone is an important issue; without those, hundreds of workers would lose their jobs."

Kempe was startled by this story and it took a while before he replied: "I agree with Eigruber, these damn Jews have to be punished. All they would do is hang jewels on their Sarah's and enrich themselves, even more with the masterpieces of our Aryan Teutonic ancestors. Therefore, I order you to fully comply with this request from your true patriotic-thinking *Gauleiter* (Region Leader)!"

The Superintendent was so crushed by Kempe's statement that he could only mumble a sort of compliance.

Early next day Kempe and Mertens, accompanied by most of the soldiers and workers from the salt mines, returned to the Steinberg Haus. The trucks had a difficult time trying to climb through the snow, and, eventually, had to abandon the attempt to reach the salt mines. Instead, the Superintendent made two caterpillar-type tractor vehicles available, which previously had been used for transporting the bombs.

Meanwhile, Mertens was busy labeling the various art pieces he had selected, still horrified, and thinking about the planned destruction of all these treasures. If only he could find a way to get the news of the bombs to Richard Johnson in Zürich, giving him time to implement some sort of Allied counter measure. Mertens' only chance was to contact Wilhelm Kuhn in Berlin, to relay the information through his neutral embassy friends to Zürich.

He not only wanted to see Erika again, but also had urgent reasons to get back to Berlin. That said, he knew action was urgently needed right at this moment.

Mertens had to find the hidden bombs and try to remove them immediately, one way or another. At an appropriate moment, he asked one of the local workers if he had any knowledge of the hidden bombs. The man, an elderly rustic character, replied in his difficult to understand mountain dialect: "I know these caverns in and out, since I have been working here for over thirty years; nothing gets by me unnoticed, including my co-workers. I'll show you where these damn bombs are located."

Walking behind the worker, Mertens passed through a number of caverns. At cleverly spaced points, single wooden crates had been placed among the rest of the art treasures. These crates were clearly marked, *"Marmor"* (marble), and under closer inspection, he noticed electrical cables connecting them. He asked the worker to open one of them, and saw the gray, rusty, shape of an enormous bomb inside. On his knees, looking into the box using his headlamp, he noticed that the hexagonal steel plug from the bomb's nose had been removed, and was lying loose on the bottom of the crate. Instead, a thick electric cable led into the bomb nose opening, obviously connected to the exposed fuse.

Although he was prepared for any Nazi insanity, Mertens was shocked at finding this planned destruction. Fortunately he was familiar with American ordnance, especially with the application of fuses. Leaning down, he grabbed the cable, and ripped it out of the bomb opening, then ordered the crate to be closed. The greatly impressed and relieved worker eagerly led Mertens to all eight-bomb locations where he repeated the defusing operations. Then he asked the mineworker to inform his colleagues, that they should remove these bombs and dispose of them outside the caves. Later, he would advise regarding the exact timing of this operation.

Returning to the Steinberg Haus, Mertens got hold of the Superintendent and asked to have a word with him in private. Both men entered the Superintendent's office, closed the door, and Mertens reported what he had done. The man was elated at the news, jumping from his chair, shaking his hands vigorously in gratitude. Mertens explained he had the workers lined up to remove the bombs, but needed a place, where these devices could be safely stored. He also explained that now would be the best opportunity to remove these bombs. They could be brought past the guards at the mine entrance as part of the treasure removal operation, and as an authorized transfer action.

"I suggest we bring the marble crates to the left side of the Steinberg Haus," the Superintendent suggested, "right next to the pile of stored firewood, where they could be covered with an old tarpaulin. Later, at a safe time, my workers can take and dispose of them somewhere in the forest." Mertens knew Kempe was busy in the morning at Alt Aussee, meeting the Nazi 'SD' boss, Dr. Ernst Kaltenbrunner, who was now in full charge of the despicable SS-Sicherheits Dienst (SS-Security Service). Kaltenbrunner took over after the monster Reinhard Heydrich had been assassinated in May of 1942, at Lidice-Czechoslovakia. Kaltenbrunner was visiting the Alpine area, where Mertens suspected he was inspecting his future hideout. He mentally made a note of this fact. Realizing Kempe and Stolz would be absent; he decided to remove the bombs during the early morning hours of the following day.

As planned, Mertens delegated all SS-soldiers to store the remaining treasure items near the mine railroad track, keeping them busy within the confines of the caverns. At the same time, he used the local workers to remove the marble crates one by one, and transport them separately to the designated woodpile location. This maneuver had to be conducted without arousing the suspicion of the SS-soldiers; therefore, Mertens concentrated his attention on them, ordering them around and keeping them from observing anything else but their own work inside the caverns.

The subterfuge and scheme worked without a hitch, and all bombs were eventually stored next to the woodpiles. The Superintendent thanked Mertens exuberantly for this audacious deed, and they parted on the friendliest terms.

The two days of arduous work using the caterpillar vehicles brought the removal and transfer of the selected art treasures to completion. Two trucks were loaded with valuables. Finally, during the first week in December of 1944, the fully loaded and assembled convoy left the beautiful Salzkammergut area in direction of Wittenberg on the Elbe River. The trip was slow and uneventful, averaging less than thirty miles per hour. They avoided all major cities on the way and used side roads. Between Plauen and Gera, the convoy halted for the first night's stop.

Early next morning the last eighty miles to Wittenberg lay ahead of them, and it was around noon when they finally arrived. At the river's small harbor, Stolz quickly located the designated motor barge named, "Malvine", and its Captain Günter Müller, together with his first-mate Hans Penner.

Müller was about fifty years old, slender, with pale blue eyes and gray hair. He was dressed in light brown corduroy pants, a double-breasted navy blue sailor jacket and a dark blue cloth cap. A shag pipe never seemed to be out of his mouth. His lingo was pure *Plattdeutsch* of the Hamburg variety.

His mate, Penner, was a simple man of about sixty, who wore a gray sweater over his dark gray bell bottom pants. He covered his head with a dark blue woolen cap. Neither Mertens nor Kempe could understand a single word that was said, when the two boatmen talked to each other in their Hamburg *Plattdeutsch.*

The barge was about sixty feet long, driven by a 200-horse power Farymann Diesel engine. Located at the stem was the pilot deckhouse, which contained adequate quarters for the Captain and cramped accommodations for the Mate. At the rear and aft of the pilot area were two more small cubicles with bunks, which were turned over to Kempe and Mertens. The space between the pilot house and the bow was an empty freight-storage hold; it was covered at deck level by a removable roof shaped lid. Part of the below forward section was immediately taken over to accommodate ten heavily armed SS-soldiers, who used their bivouac gear for their comfort. Another ten soldiers were placed directly near and below the pilothouse, where they erected a tent cover.

After Mertens, Kempe, Stolz and the SS-soldiers had stowed their gear in the designated areas, the unloading of the armored vehicles and their precious gold cargo began. Four bullion bars were packed in small wooden crates. Eventually the total reached fifteen boxes of gold each box weighing approximately eighty kilos or 176 pounds. The remaining items of coin collections, statues and small portraits were also crated in manageable box sizes to allow for two men to handle them. Finally, the entire treasure trove was moved into the middle of the barge's cargo belly, guarded from two sides by the twenty SS-soldiers. At last, fully satisfied with the transfer operation, Kempe gave orders for most of the company to leave and return to their respective units.

Kempe disappeared ashore after the SS-guards were dismissed, but returned shortly in a very excited state. From a local post office, he had telephoned his boss Adolf Eichmann, who ordered him to come to Berlin immediately for a most important conference. Mertens saw his opportunity and seized it.

"Heiner, could I go with you? As you know, I have my girl Erika still in Berlin. She's a registered nurse doing volunteer work at the Westend military hospital. We certainly could use her in an emergency for our soldiers, since we still have a dangerous trip ahead of us."

Kempe leered at Peter and said: "Don't give me this crap of caring for our soldiers. Your motive is plain to me; however I like the idea of having a young woman in our company. But you better watch out that you don't lose her on this trip. I'll leave Stolz in charge of the barge until we return. You and I can take this short trip to Berlin in my Mercedes."

It was late afternoon when they arrived in Berlin; Kempe stopped at Wilhelm Kuhn's gallery to let Mertens off. He promised to notify Kuhn what time the return trip to Wittenberg would be.

Wilhelm was delighted to see Mertens safely back from his journey.

Peter suggested another walk in the Tiergarten, where he told him about the salt mines and the insane plan to blow up the art treasures. He told him how he prevented this disaster.

Wilhelm promised to contact his embassy friend the next morning to relay the information to Richard Johnson. Now, Wilhelm brought Peter up to-date. "The German army has just started a surprise offensive from the Ardennes Forest, catching our US-troops completely by surprise. Unfortunately, the weather had been heavily overcast and foggy, rendering the Allied Air Forces blind and ineffective. Fortunately, the American First Army has slowed down the Germans near Bastogne; otherwise, they would have captured huge amounts of fuel for their tanks. But I am sure as soon as the weather clears, we will be able to force them to retreat."

"On the Russian front, the Red Army is getting ready for a huge offensive, which undoubtedly will cause the German defenses to collapse. By January 1945, the Russians will probably reach East Prussia and Poland. The question is, who will reach Berlin first, the Russians or the Americans?"

Returning to the gallery, Peter called Wolfgang Kopaneck's villa and spoke to Erika who was thrilled to hear from him, especially after he told her he would be with her very shortly.

*** *** ***

Just before Christmas, 1944, in the late morning hours, several crowded limousines arrived at the Reich's-Chancellery in Berlin's Wilhelm Straße. The city was a scene of massive destruction. Every street had more than half its buildings destroyed or gutted by bombs and fire, due to the ever-continuing air raids. On top of this devastation, a layer of snow covered shattered ruins like a funeral shroud.

Most of the officials in the limousines were either in gray or black SS-uniforms, although some were wearing the yellow-brown outfits of the Brownshirts, as displayed by Gauleiters and other party officials. Eichmann, accompanied by Kempe, was part of the group following their leader, the man in charge of *Gestapo*, SS and police forces, Heinrich Himmler.

The entire party hierarchy was assembled in the large tile-covered Reich's Chancellery courtyard. A number of steps led into the main entrance, guarded by two large sculptures of naked heroic men, clutching torches in their outstretched hands. At this point, four SS-guards stopped and checked the credentials of the approaching officials.

Eventually, the invited group assembled in the large festival hall, where previously all so-called diplomatic affairs had taken place. Whispering to each other, the old party comrades exchanged latest information and rumors. A somber atmosphere pervaded the entire group. They had waited for about half an hour, when quite suddenly their *Führer* Adolf Hitler entered the hall. Enthusiastically the crowd shouted: "Heil Hitler" and "Sieg Heil" over and over again.

Hitler looked worn, broken, and tired in his gray army coat and black trousers. He was slouched forward, then with a bent arm and raised hand he responded to the crowd. He approached the microphone, slowly positioned himself to speak. Taking his time, waiting for complete silence, he finally began, speaking in an old man's voice.

"My dear comrades and party members. I have asked you here to bring you up to date regarding latest events. Also, my future outlook will affect all of us. We are presently engaged in a struggle against the world of Judaism, whose only goal is to destroy Germany and enslave our people. I promise we will not stand idly by, doing nothing, letting this monstrous plot succeed. I have sworn to fight this vermin of humanity to my last breath and beyond, even if we should ever lose this war. However, I am still confident our miracle weapons will turn this struggle in our favor. At the moment our armed forces have lashed out at the enemy from the Ardennes Forest and soon we will reach the English Channel, cutting the enemy in half. We shall throw the servants of Judaism into the ocean, and this time, I will not show leniency, like before at Dunkerque."

At this point of the speech the crowd roared with almost hysterical response, shouting over and over again: "Sieg Heil, Sieg Heil!"

Then Hitler continued: "One of our secret weapons is a newly-developed U-Boat, driven by the Professor Walter propulsion system. We currently have four of these submarines available; one is still being tested near Danzig in the Baltic Sea. These vessels have the capability of staying submerged for many days. But the most amazing fact is their underwater speed; they can out-run just about every enemy surface vessel." This statement resulted in resounding exclamations and expressions of awe.

"Of course, we have many more miracle weapons of which the V-1 and V-2 are already in use. These rockets eventually will be powerful enough to reach New York. Our scientists are working on a super bomb, only one of which will be able to destroy cities as big as London or New York."

"At this moment we are fighting a time element against the latest Allied landings, together with a military threat from the east. Therefore, should the enemy ever reach the Reichs-Haupt-Stadt Berlin, the following plan shall be activated:

"I have prepared a list of party members, who I believe are most suited to carry the idea of National Socialism throughout the world, who would continue my fight against·'International Jewry'. My staff officer Martin Bormann will distribute this list to you. It contains the names of forty members. A group of ten each will board the four U-Boats as passengers to destinations in South America. The submarines will also be loaded with gold bars and valuables, making it possible to establish and finance a successful organization either in Argentina, Paraguay or Brazil. The departure date and time will be determined by me personally."

On hearing this news the listeners were stunned, each man wondering if his name was on the list.

"Comrades!" Hitler continued. "Unfortunately, only four of these vessels are available, otherwise every one of you would have been chosen for this important assignment. Every one of us must be responsible to carry on the ideology of National Socialism with its fight against 'International Jewry'. If necessary, we shall activate 'Werewolf' groups who must fight, if essential, to the last man and woman of our nation. In this spirit I shall say farewell to all of you. But I demand that we intensify our fight, until we can establish enough time for our miracle weapons to be used. Now, Martin Bormann will call the names from the selected list, Sieg Heil!"

Before Hitler could leave, several members of the crowd shouted: "Don't stay in Berlin, use one of the submarines and guide us into the future."

To these remarks Hitler replied:

"I have decided to stay in Berlin and lead Germany, if necessary, to its final end. This is my destiny, but yours shall be to carry on the fight for 'National Socialism', Sieg Heil!"

After Hitler had left, Bormann, a stocky, brutish bully in a brown uniform stepped to the microphone, reading the names from the *Führer's* list. Each time a person was called and identified he stepped forward and aside, creating a separate group. Kempe's name was called and, elated, he strode over to the selected circle of Nazi-fanatics. In a few moments, the chosen forty received identification documents certifying their assignments.

On the way back to the office, Eichmann congratulated Kempe on this privileged assignment, emphasizing that he had recommended him highly to Himmler and the *Führer*. He hinted that he himself might still join the special group by escaping later to Argentina, one way or another. However, at the moment, he had declined this possibility in favor of handling the most important action of the 'Final Solution', meaning the extermination of the last Jews in Europe.

Kempe thanked Eichmann for his confidence in him. After returning to his office he sorted his papers and locked his desk, then telephoned Wilhelm Kuhn. When he learned Mertens was at Kopaneck's villa, he called him there and ordered him to get ready with Erika for the trip back to Wittenberg. Then he went to his Mercedes and drove to Wolfgang Kopaneck's house.

At last Kempe knew why he had collected all these treasures from the salt mines. Now he could guess the barge transport destination, namely, the submarine base near the city of Kiel. He also understood why Eichmann had sent him there beforehand to make him familiar with these U-Boats.

Mertens wrote a hasty note to his friend Wolfgang Kopaneck, who at this time was already at his office in Reinickendorf. He told him about his sudden departure, and thanked him for his hospitality, promising he would look him up at a future date.

Kempe, who had never met Erika, was highly impressed by her Nordic beauty.

He was neither aware of her half-Jewish background nor that she was the daughter of one of his many murder victims. He envied Mertens, and the urge provoked him to try everything and anything to take her away from him. He displayed all the charm he could muster. Gallantly, he placed Erika in the front seat of his automobile, compelling Mertens to sit in the back. Two hours later they were in Wittenberg boarding the barge "Malvine"

CHAPTER 16

After Kempe met Erika, Mertens noticed the SS-man's attitude toward him changed abruptly. During the trip to Wittenberg, he sensed that Kempe was completely ignoring him, concentrating all of his attention on Erika. Mertens suddenly realized an enormous problem had developed, and he frantically searched for a satisfactory solution. Killing Kempe would be premature, because Mertens assignment was to get to the submarines, and Kempe was the only person who could lead him there. On the other hand, he could not allow Kempe to seduce Erika.

Stepping aboard the barge "Malvine," Stolz reported everything under control and ready for the trip north. Kempe ordered the Sergeant to take the Mercedes, drive to Glückstadt, south of the Kiel Kanal, and wait for the barge's arrival. After Stolz left, he called Mertens to join him in the pilothouse. The Captain and his Mate were busy checking the engine room.

Kempe thanked Mertens for his assistance on the trip, handed him an envelope and said, "Peter, you have completed your assignment very well. As I said before, here is the promised five thousand *Reichsmark*. You can leave now and take care of your own private affairs. Of course, I would want Erika to stay with us here on the barge, because your suggestion of having a trained nurse on board was excellent. So I suggest you pack your belongings and leave as soon as possible."

"I shall do as you wish Heiner; however, Erika is my fiancée and we intend to get married. Therefore, I expect you'll act accordingly, and you'll treat her like a gentleman."

Kempe's eyes hardened and in the half light Peter detected a lecherous leer: "Of course, Peter," he said cordially. "I shall take good care of her. No harm will come to her."

He then shook Peter's hand and left the barge heading for the post office, where he planned to report to Eichmann that he was ready to continue his trip north.

Peter immediately found Erika and told her of these new developments, suggesting she should gather her belongings and that they both should just leave, at once. Erika thought this matter over carefully, and decided that if their mission required it, she would take her chances with Kempe.

While they were debating their options. Kempe suddenly reappeared on board. He just had spoken to Eichmann. In a surprisingly conciliatory manner, he said to Peter: "Our plans have changed. Please stay on board, as your services will again be needed. We expect another shipment of art treasurers to arrive near the city of Kiel, and we need you to sort out the most valuable items. I am authorized to pay you another three thousand *Reichsmark* after completion of this assignment. Of course, there is no choice; you must accept this important directive."

Peter was relieved at not having to leave Erika, but felt suspicious as to what the Nazis wanted with all of these valuables. He couldn't help concluding, that these treasures were being prepared for shipment out of the country, but by whom, and most of all, why?

"Heiner!" he said: "Of course I'll be more than happy to accept the money and follow your new instructions."

Erika was given the Mate's cabin, while Peter had to share accommodation with the Captain, whose cabin contained two bunks. The departure was set for early next morning, and the Captain invited everyone to a surprisingly delicious dinner in the pilot house. The meal had been expertly prepared by the Mate, who had once served as a chef on board various luxury liners, before the war.

Both seamen were great characters with grand senses of humor and a repertoire of salty jokes. The Captain told his guests, with a twinkle in his eye, that he had once met a communist.

"I asked this man, would he give me one of his homes if he had two of them? Of course I would, he said. Then I asked him, would he give me one of his bicycles if he had two of them? Again he answered; of course, you could have one of them. Finally, I asked him could I have one of his duffel bags if he had two of them? The communist replied: "Hell no!" Surprised I asked why not, to which he answered: "Because I happen to have two duffel bags."

Kempe thought the story was hilarious, and roared with loud laughter. The mate concluded the evening by playing a number of sea chanteys on his accordion. He sang in a deep rough voice a song about a seaman longing to return to his beloved girl in the harbor of Hamburg.

Early next morning, the weather was misty and cool. The Mate, on orders from his Captain, removed the lines from the mooring posts and the barge engine moved the vessel into mid-channel of the Elbe River. . The soldiers made themselves comfortable on deck and played cards. The trip was monotonous; the scenery on both sides of the river was bordered by flat meadows, interrupted occasionally by a small village. The only diversion of the trip was when the barge had to enter a river lock.

After the lock gates had been closed, propelling the water through the opened floodgates ahead of them, the lock gate level was lowered. The first part of the trip consisted of many river turns. Shortly before evening, they reached a point south of the city of Magdeburg, where they moored at the river's edge.

Again, the mate produced a good evening meal with two bottles of French red wine. Everyone listened to the Captain's hilarious stories, and Erika helped clear the table and wash the dishes. Above them, hundreds of allied planes proceeded to their pre-assigned targets, accompanied by the thunder of the German FLAK artillery.

By evening on the next day they reached the outskirts of Wittenberge.

On the evening of the third day, they arrived at the town of Lauenburg, located about forty kilometers south of Hamburg. It was Christmas Eve, and the Mate produced a small spruce tree that he had decorated with tinsel and candles. Safely moored, accompanied by the Mate's accordion, the soldiers, crew, and passengers sang holiday songs and everyone fell into a sentimental mood. *"Stille Nacht, Heilige Nacht"* or "Silent Night Holy Night", sounded solemnly from the barge.

Half-drunk, Kempe tried several times to force his attentions on Erika, who continued to evade his persistent advances. Peter struggled to hold his temper, still remembering Kempe's rape attempt in Paris. When Kempe tried to force a kiss from Erika, Peter knew he had to take drastic measures. Knowing the German mentality too well, he shouted at the drunken oaf, saying, "You are not behaving in accordance with the ethics of a German officer," he shouted. "Get hold of yourself, or you will not be fit to wear your uniform. You're setting an ugly example for your men!"

These remarks cut deeply into Kempe's ego. He slowly turned toward Peter, stared at him with hate-filled eyes, then walked outside and disappeared from the party.

Early on Christmas morning, Kempe avoided Peter and Erika, and spent most of his time with his men. The barge continued its journey, taking the left branch of the Elbe leading into the Süder-Elbe arm of the river, bypassing Hamburg. Looking north, the silhouette of that majestic city could be seen even though most buildings were either destroyed or nothing but burned-out shells. Passing through the Köhl-Fleet section they entered again the main stream of the Elbe. Subsequently, they headed west and away from the Allied target of Hamburg. During the dark hours of evening, they reached their goal, the small town of Glückstadt on the east shore of the winding river. Stolz was already waiting for them with the Mercedes.

The next morning just after the post office opened, Kempe left the barge to place another telephone call to Berlin. He was to receive his instructions for the final destination of the expedition. Coming back, he immediately ordered Stolz and Peter to meet him alone in the Mercedes for a conference. There he instructed them about their next movements: "Stolz, you and Peter will drive straight to the Kiel harbor pier, where the high speed Schnellboot S-19 is moored. You'll leave the Mercedes at the pier to be picked up by me at a later date; I have my own set of keys. The skipper of the boat is instructed to take you both on board, and move to the Baltic port of Gothenhafen, the former Polish harbor of Gdynia near Danzig, where the passenger ship, 'Wilhelm Gustloff', is docked. The repair work on the ship's bomb damage should be almost complete. She is under orders to pick up German civilian refugees, mostly women and children, including wounded soldiers and have them evacuated to Stettin and Lübeck.

''Your responsibility will be to receive twenty-three crates of valuables and have them safely stored on board the Gustloff. Once the ship has reached Lübeck, these crates have to be transported to the submarine base in Eckernförde. Peter's job at Gothenhafen is to inspect the cargo and shipment lists, together with the bill of lading. The return trip of the Gustloff shall be via Stettin. The departure is scheduled to take place toward the end of January, 1945.

"Peter, I have received special instructions for you. You're ordered not to return with the Wilhelm Gustloff, as only Stolz is needed there. You'll visit the isle of Hela near Danzig, where one of our latest submarines is concluding its last tests. You are to report to submarine Captain Hilmar Volker, who has been advised of your arrival. You are to go on board the U-Boat, inspect its storage capacity for the art treasures, and plan with the Captain where and how to store them. You might just as well return with the U-Boat to Eckernförde sometime in February, where we all will meet again.

"Do you have any questions? No? Okay, get your gear and get ready to leave."

Returning to the barge, Kempe instructed the Captain to continue the trip to the entrance of the Kiel Kanal, a waterway connecting the Elbe River and North Sea access with the Baltic Sea.

Before leaving, Peter was anxious to learn about the outcome of the Ardennes offensive by the German army. He also wondered what the situation was in the east. Had the Russians started their expected offensive over the Vistula toward Warsaw? Captain Günther Müller was the right person to ask. During the trip Peter had become quite friendly with him, and it took no time at all to realize, that this world traveler was an astute judge of Germany's situation, and her fate in the war. Where the skipper got his facts, Peter did not want to pursue; but the Captain told him the Ardennes offensive was over, that soon after the weather had cleared, the Allied air force had regained control over the sky. In the east the Russians were still preparing for the soon expected all-out offensive against Germany itself.

Feeling heartsick about leaving Erika behind with this dangerous SS-man was depressing to Peter. But he saw no other choice and said farewell to Erika. Locked in each other's arms, they whispered their hopes that soon they would be happily united again in Eckernförde.

CHAPTER 17

On a clear chilly day with temperatures around freezing, the barge "Malvine" passed a major lock, and moved slowly along the canal toward the Baltic harbor of Kiel. All day long, brooding Kempe had avoided Erika, spending most of his time in his cabin, or speaking to his men. He had become deeply depressed, realizing for the first time that the war might be lost, and the glorious future he always dreamed of would probably never come to pass.

Fortunately, the *Führer* had provided him with a special future, however inglorious, by allowing him to leave Germany in a secret U-Boat. This handsome Peter Mertens was another depressing matter. Mertens was getting on his nerves: "What right did he have to tell me, how a good German officer should behave? To hell with Mertens and his slut! She was just another woman who had to be shown how a representative of the master race would rule and what her obligations were to an officer of the Third Reich. She should feel honored that he desired her!"

Gleefully, Kempe remembered the times when he had been ordered to visit the *Lebensborn*, or 'life procreating' establishments, in various parts of Germany. He was one of the specially chosen SS-officers, ordered by Heinrich Himmler, to mate like a stud with some completely unknown, but carefully chosen, Aryan girl for the purpose of fathering children for Germany and the *Führer*. Oh, it felt so good to have sex so often, spread his seeds and have his genes live on, carried by these emotional, almost hysterical women, who were duped to believe that any boy they bore would be a genius reflection of Adolf Hitler himself.

Later toward evening, on the outskirts of Kiel, the barge moored near the large lock leading to the harbor, from which the Baltic Sea could be reached. Once again, the mate prepared a sumptuous meal, and adding a few bottles of wine. During the dinner, Kempe couldn't keep his eyes off Erika. He decided to try to get her drunk, offering one glass of wine after another. Erika suspected what he was up to and avoided this obvious ploy by excusing herself, saying she was tired, had a headache, and wanted to return to her cabin. As Erika stood to leave, Kempe offered to escort her, but in spite of a polite refusal, he followed her to the cabin door. Without a word, he jostled them both inside the room, grabbed Erika around the waist and threw her on the bunk. Suddenly, wailing air raid sirens started their terrifying chorus all over Kiel, followed almost simultaneously by the furious barking of FLAK artillery in angry concert. The barge shuddered from the overhead drone of hundreds of British aircraft engines as their first bombs exploded in the water and around the harbor area.

Startled, Kempe cursed this interruption, but decided to ignore the air raid and continue his rape-like attack. To hell with the bombs falling in the harbor or the huge blasts and deafening air concussions. In the midst of this raging madness, a number of frantic shouts for medical help could be heard from the barge's deck.

They had a sobering effect on Kempe, who promptly stopped his rape attack. He turned and hurriedly left the cabin, running toward the deck area, followed by Erika. Both arrived on deck to find a group of soldiers leaning over two wounded comrades. They made space for her to examine the air raid victims. She established immediately that one of the two men was already dead. A fragment of FLAK shrapnel had struck him in the neck killing him instantly. The other victim had also been hit in the shoulder by shrapnel, and he was in great pain. Erika dressed the wound as well as possible, gave him an injection of morphine to lessen the pain, and advised officials to take him as soon as possible to a military hospital.

Due to the air attack and its damage to the area, they had to wait a long time before an ambulance arrived and removed the body and wounded SS-soldier. Kempe's amorous inclinations had been quenched by these events. He sank again into depression, and withdrew into his cabin with a bottle of cognac. Erika returned to her cabin, locked her door, and hoped for an undisturbed nights sleep.

Early next morning as soon as the large lock to the Baltic opened, the Captain moved the barge to get out of Kiel, a dangerous target area constantly attacked by the Royal Air Force. The sight around the harbor was awesome; the bomb damage had twisted cranes and steel structures.

Along the piers, a number of storage sheds still burned or smoldered in the early morning haze.

The weather became ugly; sleet fell, obscuring the view ahead of the barge. Nevertheless, the Captain steered the vessel toward the open sea, reasoning that since the wind came from shore, a westerly direction, the land would act as a protective shield against the wind. Passing the Bülker light house, the barge kept close to shore, then turning north-west toward the bay inlet of Eckernförde. The trip took about four hours and after passing through the Eckernförde harbor inlet, they reached their final destination, the H_2O_2 U-Boat bunker installations.

Kempe went ashore and immediately reported their arrival to Eichmann by telephone. Erika hurriedly left the barge searching for a place to stay. She found a small naval hospital where she was eagerly welcomed, and given a room in staff quarters, which she had to share with another young nurse.

Erika told Kempe about her new quarters and job assignment, vowing to stay as far away from him as possible. Kempe was furious and at first, threatened her with punishment for having left his command without his permission. Finally, he consented, not wanting to be further embarrassed in front of his men who, of course, where fully aware of his nagging infatuation with Erika.

*** *** ***

Stolz drove, while Mertens sat in the front passenger seat, enjoying the scenery of the trip from *Glueckstadt* to Kiel. They passed through a magnificent area the natives called, *Holstein*-Switzerland. The beautiful landscape was studded with clear lakes, meadows and tall forests. After a journey of about thirty kilometers, they reached a town called *Itzehoe*, where both had lunch at the quaint city square inn.

Stolz was a taciturn fellow who spoke barely a word, except in connection with the present assignment, or when checking with Mertens, who had the road map on his lap, for directions to Kiel. Again, Stolz thought he had met Mertens before, but could not place him in his memory. It just about drove him crazy.

Early in the afternoon they reached Kiel, where they quickly found the enormous harbor, with its submarine bunker installation, ferry piers to Scandinavia, and Baltic steamer moorings . At the submarine bunker they found the torpedo speedboat "S-19". The skipper had been expecting them. After the Mercedes was safely parked, the two men boarded the boat to be shown below to their quarters, consisting of two hammocks. The crew included fourteen men, plus their skipper and his first officer.

The boat was painted dark gray, built with heavy timber framework, with a hull comprised of multiple layers of plywood. Forward on deck were two torpedo tubes, one on either side. A massive, double-barreled 20-mm machine gun protruded from the cockpit area toward the rear. The radio mast and moveable spotlight installation faced forward.

Losing no time, the skipper gave orders to cast off and the thirty-five meter long boat headed cautiously out of the harbor toward the open sea. The sky was gray, overcast with gusty westerly winds. Because it looked like it would snow at any moment, the crew was heavily dressed in sweaters and dark gray weather gear.

After passing the last cardinal harbor buoy, the skipper opened the throttle of the 500 horse power gasoline engines, which responded with an awesome roar. They drove the planing hull out of the water, skipping with heavy bounces over the somewhat turbulent sea. Every so often a large wave would crash over the bow sending a deluge of salt water and spray over the cockpit area that was protected only by the navigator's super-structure. The skipper told Mertens they were presently cruising at about 40 to 50 kilometers per hour, but could develop a maximum speed of over 70 kilometers, depending on the prevailing wave conditions.

Mertens and Stolz went below into the cabin, where it was more comfortable and a lot warmer. The continuous bouncing over the swells of the sea became so uncomfortable that they eventually just grabbed their hammocks, climbed in and stretched out, letting the supporting lines absorb the bone-jarring shocks. Surprisingly, nobody became seasick.

At 5 p.m., they passed the isle of *Fehmarn,* and then continued in an easterly direction toward the island of Rügen. Wisely, the skipper avoided a stopover at any of the larger harbor towns, not wishing to be caught by a surprise air raid. Just before reaching Rügen he reduced speed, then cautiously headed southeast through a narrow passage into the small harbor of Stralsund. There they were to stay safely overnight, moored alongside one of the piers.

Early next morning they left Stralsund, moving slowly in a southeasterly direction, passing underneath the mainland causeway that connects the island of Rugen. Then they entered the so-called Bodden, a narrow straight that would lead them into the open Baltic Sea again. They still had to cover about 350 kilometers in order to reach Gothenhafen near Danzig. The weather had settled down and was calm, so the skipper increased the speed to 70 kilometers per hour, passing through the Greifswalder Bodden and the northern most tip of Peenemünde. Suddenly, an enormous ear-piercing roar drowned out the noise of the boat's engines.

South of them, they saw a fiery trail rising from the base of the island. In utter amazement they noticed a missile shaped rocket moving overhead and disappearing in the west. This was the first time Mertens saw Stolz get excited.

"I knew it! I knew it!" he shouted at the top of his lungs. "The *Führer* is releasing our miracle weapons, the war will turn in our favor and all those damned Jews still left in the world, will get what's coming to them!"

The skipper and his crew reacted much more calmly, because they had passed this point a few times before and had already witnessed the launching of the V-2 rockets. Mertens learned that the letter 'V stood for the German word *Vergeltung*, or 'Retaliation'.

Just about noon the "S-19" reached Danzig's bay where, heading south, they passed a narrow peninsula. They saw at its most southern point, the fishing village of Hela. From there they headed straight west into Gothenhafen's harbor.

Entering the port they recognized the huge ivory hull of the liner 'Wilhelm Gustloff," which was moored alongside a pier. Not far from there the skipper found the berth for his S-boat. A new group of passengers, consisting of naval officers and one high- ranking Nazi, in a brown uniform, with his wife and two children, were already waiting to be taken aboard for the return trip.

Mertens thanked the skipper for a most interesting journey, and then he and Stolz left the boat, heading for the gangway of the "Wilhelm Gustloff". There was, in fact, more than one gangway connected to the ship, and a steady flow of workers moved on and off the vessel. Getting closer, Mertens noticed considerable damage to the liner's decks, obviously caused by enemy bombs.

Two armed military policemen were guarding the main gangway, while another couple of soldiers patrolled the pier area. After the MPs had checked their papers, Mertens and Stolz boarded the liner, making their way upward to the bridge area. The Captain was absent, but his first officer had been instructed to meet them.

Stolz was assigned a double bunk cabin for the upcoming trip, and Mertens for the time being, shared one of the two available bunks. It was agreed that Mertens should go below and get an idea of the limited storage area to be used, ultimately transporting twenty-three treasure crates. The Gustloff was not at all designed for transporting freight; she was built strictly to accommodate vacationing travelers. The First Officer explained the design of the liner, and the reason for the ongoing construction activities.

"The 25,487 gross register ton ocean liner was built in 1939 by 'Blohm and Voss' in Hamburg, the same shipyard which built the battleship 'Bismark'. She could accommodate 1,465 passengers, is 195 meters long and 23.50 meters wide. She has a draft of 14.50 meters, which is the reason the repair work couldn't be done in the harbor of Danzig with its fairly shallow waters.

"Please tell me about the ongoing repair work," asked Mertens.

"The work became necessary due to damage received from bombings, while moored in the harbor of Konigsberg in East Prussia. The ship's power plant consists of a 9,500 horse power diesel system, having four 8 cylinder, two stroke engines, two of which are connected to each of the two propeller shafts. The maximum cruising speed is rated at about 15. 5 knots."

"I imagine the ship's speed will protect her from enemy submarines, which aren't capable of more than 1 0 knots," interjected Peter.

The Officer agreed with him, but didn't want to be interrupted in his lecture and continued.

"You will notice the deck areas are free of obstructions, such as winches and ventilators, etc. providing ample space for the passengers to spread out and enjoy the travel. Professor Waldemar Brinkmann of Munich designed the interior, with its beautiful social and music hall, dance floor, library, gymnasium and tiled swimming pool. Let me take you through the ship; then you can decide what you need for your storage area."

Having inspected the liner, Mertens returned to. the cabin, where he decided the best storage area for the crates was the empty swimming pool. However, he wanted to wait, before making the final determination, until he had seen the 23 crates, which were presently kept at a waterfront warehouse in the city of Danzig.

Early next morning and after a good night's sleep aboard the Gustloff, Mertens and Stolz caught a tender boat from Gothenhafen to Danzig. Entering the harbor, called *Neu Fahrwasser*, they disembarked at the pier, and took a short walk to the designated storage building. The facility was guarded by five SS-soldiers, who carefully checked both of their credentials. Then they met the local museum's curator, who was in charge of the stored items.

The building was a typical hanseatic framework structure, with a steep roof and a hoist beam protruding from the upper front gable of the facade. The curator led the visitors up old worn, wooden stairs to the top floor, where all 23 crates were neatly stacked. After he opened the storage area outer double doors, the access to a hoist became apparent. Now, enough light penetrated the room for Mertens to view the freight list with its itemized contents. Many of the items were Russian icons, including more gold bars, also some Polish gold and silver church goblets, all of which were obviously stolen from the war zones.

Mertens instructed the curator that on January 19th prior to her departure, he must make arrangements to have the 23 crates transferred to the 'Wilhelm Gustloff', advising that the sailing date be set for the end of January 1945. Through Stolz, he informed the SS-soldiers to sequester the tender boat as a special transport vessel, taking the 23 crates of freight on January 19th from Danzig to the pier in Gothenhafen.

Returning to Gothenhafen at noon, Mertens located another tender, which took him and Stolz east across the bay to the fishing village of Hela, where the Walter submarine G-17 was moored alongside the pier. Rafted against the U-Boat and toward the seaside were two old fishing trawlers, probably serving as a protective barrier against outside attacks. Two marine soldiers with machine pistols guarded the submarine. Mertens asked to see the Captain who came from the conning tower to meet them.

"I am Captain Hilmar Volker, what can I do for you?"

"This is *Schaarführer* Horst Stolz and I am Peter Mertens! I am ordered to report to you under the code name 'Golden Eagle'!"

The Captain was a youthful handsome, blond fellow with a well-groomed beard. He wore a white U-Boat officer's cap tilted to one side, and prominently around his neck, hung the '*Ritterkreuz*' or Knights-Cross, indicating he had proven himself as a hero of Germany. He smiled broadly and invited Mertens to come aboard, ordering the SS-man Stolz to wait outside on the pier.

"I have heard about your visit, Herr Mertens, but no other person is authorized to come aboard; please follow me, alone!"

Mertens, who had never before been aboard a submarine, was surprised how small and narrow the boat was, and how compactly she was laid out. Climbing down into the belly of the vessel he had an uncomfortable feeling of confinement. The control area was just as he remembered from some of the movies he had seen, while the forward section was arranged solely as a torpedo station, including eight berths for crew members. The very cramped Captain's quarters was next to the control area, together with a berth for the First Officer. The aft section beyond the control station was dedicated entirely to the propulsion machinery.

Mertens' initial impressions were, that it would be impossible to store any of the treasures in the boat, unless they could reduce the number of crew members, and eliminate all torpedoes. He realized that no more than five or six passengers could be accommodated and even then, only if the crew consisted of the Captain, the First Officer and maybe three more crew members. In addition to the treasures, the boat had to take on board food provisions for a trip of approximately twenty-five days, for eleven people.

The powerful, swift, hydrogen peroxide engines could be used only within hostile waters, since the hydrogen peroxide fuel capacity was limited. Consequently, most of the travel would be by conventional diesel power. Mertens estimated that at least 2,000 gallons of diesel fuel would be needed to propel the boat to the nearest refueling port located at the northern most point of Brazil. Undoubtedly, the boat could accommodate the needed diesel fuel.

While Mertens was thinking about these various technical data, the Captain invited him to be seated at the chart table. One of the crew members brought a pot of hot, real coffee and the two men had a chance to get acquainted.

"Do you know this boat should carry ten passengers on a trip to South America?" Peter said. "Looking at your vessel, she certainly can't manage more than six, assuming it's possible to eliminate five men of your crew."

"What do you expect from these ridiculous amateurs, most of these party phonies are too stupid to count to three! The idea of helping these people escape, should the war be lost, is totally unacceptable, and to spread their party philosophy from South America all over the world is revolting. What do you think?"

Peter was astonished at the Captain's frankness. He obviously had no love for the Nazis.

"I am here only to do a job, but I fully sympathize with your statement.

It's obvious the party leadership realizes their ship is sinking and the war is lost. "

Mertens statement was like pouring fuel onto a fire and Volker burst into an emotional reply:

''This war has been a crime against Germany and the human-race; look at the destruction of our cities, the inhuman treatment of the Jews, Poles and other minorities. Every German, at least subconsciously, realizes the burden; he eventually must carry for the crimes perpetrated by these mad fanatics under Hitler. The motto you hear every day, 'Children, enjöy the war, the peace will be devastating!' is all too true. It explains why the German army is still fighting, even more so after the Allies have demanded an unconditional surrender from Germany."

Mertens was greatly surprised to hear these words coming from a German naval officer but he was delighted to have met this brave seaman. His thinking and personality had an enormous effect on him, and so he decided to ask some pertinent questions regarding the submarine's technical data.

"Captain, tell me, how much fuel does this vessel carry and what other performance data can you give me?"

'Well, I will tell you, the diesel fuel capacity is 20.2 tons (5,336 gallons), the hydrogen peroxide capacity is 55 tons (14,531 gallons). Our diesel cruising speed is 8.5 knots (almost ten miles per hour), while our under water hydrogen peroxide (H_2O_2) cruising speed is about 25 knots (almost 29 miles per hour) and is only used for limited periods during times of emergency or danger. Our surface cruising distance, with diesel power is about 3,000 nautical miles (3,450 miles). When using the H_2O_2 power plant in a submerged state, we can cover 125 nautical miles; hopefully enough capacity to avoid enemy attacks. Clearly we have the capacity to reach South America.

"How long will it take to conclude your test program?" asked Mertens.

"In one week I will have finished our local test program, which so far has been very successful! I also understand you are ordered aboard for the trip to Eckernförde on January 20th. In the meantime I'll be busy, and you'll undoubtedly have to attend to your own assignments. I'll expect you aboard the night of January 19th at 18:00, so we can leave early the next morning. Fortunately, there is very little tidal movement in the Baltic Sea, so we are free to leave at any time!" This concluded the Captains explanation.

At this point Peter shook hands with the Captain, departing without the Hitler salute. Stolz was waiting for the shore return trip, while smoking a cigarette.

Mertens spent most of his time between Gothenhafen and Danzig and eventually, by the afternoon of January 19th, all of the 23 crates were safely stored inside the Gustloff's empty swimming pool. More and more civilians were arriving from the east, including hundreds of wounded Wehrmacht soldiers, who received priority for boarding the liner and its newly arranged medical facilities. Masses of refugees, women, and children, were slated to embark for transportation west to Stettin and Lübeck.

The liner's First Officer had been ordered to make room for over five thousand passengers. The ship engines were only partly repaired for the trip. Regardless, the departure date was set for the early morning of January 30th. The Russian army was closing in fast. Thousands of civilian refugees were massing along the waterfront hoping to board the former luxury liner. Any departure delay would jeopardize the successful escape from the approaching hordes of Russian soldiers, seeking their cruel revenge.

CHAPTER 18

During the evening of January 19th, having completed the transfer of the treasure, Mertens arrived at the harbor in Hela to board the U-Boat G17. Captain Volker welcomed him with a cordial navy salute, while First Officer Gerhard Lange greeted him with an enthusiastic "Heil Hitler." This was the first time Mertens had met Volker's number one, as the First Officer had just returned from Danzig, organizing supplies for the trip.

The crew consisted of eight seasoned sub-mariners whose quarters were in the torpedo room bow section. Mertens had to make do with a hammock, temporarily slung in a corner near the engine room and galley.

One of the crewmembers prepared a meal for the Captain and the First Officer, while they were both occupied plotting the course for the following day's sea voyage. When they had completed the task, the Captain ordered one meal to be added, inviting Mertens to join them for dinner.

Gerhard Lange, the First Officer, came from Schneidemühl West-Prussia and spoke German with a pronounced broad drawl intonation, which left a somewhat bumpkin impression. Lange was a young man of twenty-five, with a clean-reddish face, blondish, but sparse hair, and light blue eyes. Mertens immediately recognized in him a devout Nazi supporter, who probably still believed the war could be won through the use of the *Führer's* upcoming miracle weapons. He also noticed Lange's contempt for his Captain, whom he openly criticized for not putting his trust and faith totally in Adolf Hitler's genius. Obviously, the relationship between these two officers was strained and becoming more so. Volker, an outspoken man, was having great difficulty in concealing his true feelings about the Nazis and their fanatic supporters.

About eight o'clock next morning, at the first light of day the U-Boat left the Hela pier heading north into the Baltic Sea, plowing through the water's turbulent surface. The boat was driven on by the continuous hammering of the tireless diesel engines. At the same time the H_2O_2 turbine boiler system was being primed to be ready, should it become necessary, to dive and switch over to the powerful underwater drive. The weather was heavily overcast, gray and unfriendly with a strong westerly wind churning the cresting sea into waves of ten feet.

Dressed in warm sweaters and waterproof heavy weather gear, the Captain and First Officer suggested Mertens to use spare oilies and join them in the cramped conning tower. Occasionally, huge white-crested waves crashed against the port side of the U-Boat, spraying a salty downpour over the men in their lofty position. Having left the last traces of land behind, the boat turned into the wind, heading due west. Through a steady cycle of pounding motions, the bow dug into the waves sending continuous showers of icy cold seawater over the exposed men.

Mertens thoroughly enjoyed this wild ride through the rough seas. It made him feel part of nature and the overwhelming elements of the ocean. Suddenly, an alarm bell, from the boat's control compartment, sounded in the conning tower. The Captain ordered everyone below. Being the last man down, he resealed the tower hatch. Instantly, the crew was at their stations, ready for diving. Within seconds the ballast tanks received the swift entering seawater. The boat began to submerge into the depths of the sea. Immediately, a hissing sound from the engine room signaled the activation of the turbine boiler system. A slight vibration of the boat's hull indicated that the H_2O_2 propulsion system was on line, and had taken over the thrusting power.

The sonar crew member at his post had received the pinging echo of another submarine in the area. The U-Boat G-17 was unarmed, and, in accordance with her test assignment, carried no torpedoes. It became necessary to resort to evasive maneuvers, relying entirely on the vessel's underwater speed for protection.

In a tense voice, the sonar operator reported the sound of a torpedo heading in their direction. The Captain ordered full speed ahead, and the U-Boat surged forward with a jolt of greatly increased speed. After a few suspenseful moments, the sonar operator announced in a relieved voice, that the torpedo had passed the vessels stern, barely missing her. Thanks to the boat's revolutionary underwater power system, a major disaster was avoided.

Captain Volker praised the sonar operator for his diligence, and said.

"This must have been a Russian submarine, as no other allied vessel would dare to enter the Baltic Sea through the narrow inlet between Denmark and Sweden. We must immediately notify the Captain of the 'Wilhelm Gustloff,' warning him of this possible danger!"

Captain Volker decided to maintain a submerged speed of twenty-five knots for a period of two hours before reducing the velocity and bringing the U-Boat to periscope depth. Finding nothing by scanning the horizon and checking the sonar, he ordered the vessel to surface, and revert back again to the diesel engines. Hydrogen peroxide was valuable, and was to be used only in times of emergency.

The rest of the trip was uneventful, and the U-Boat arrived in Eckernförde, entering the giant submarine bunker installation during the early morning hours of January 22nd, 1945. She rafted alongside three more submarines of the same type, to which she was securely moored.

Leaving only a skeleton crew aboard, the two officers, Mertens and several sailors stepped ashore. Mertens found Kempe in the base cafeteria with two civilians dressed in the typical leather jackets of the infamous *Gestapo*. After having been introduced to them by Kempe, Mertens reported news of his latest assignment.

Kempe's mood was morose and rather unfriendly, but he told Mertens to talk freely since the two civilians were friends. Mertens gave a full account of the U-Boat's journey and the entire trip, explaining the storage arrangements of the treasure in the swimming pool of the 'Wilhelm Gustloff.' Kempe seemed satisfied, especially after hearing Mertens' report that Stolz was now on board the liner.

"I want you to remain here with me until the 'Wilhelm Gustloff' has arrived in Lübeck," Kempe said to Mertens. "You and Stolz will then be responsible for transferring the treasure to the Eckernförde submarine base, storing it alongside all of the other valuables. The Gustloff is expected to arrive on the morning of February 2nd. In the meantime, I want you and Captain Volker to show me our accommodations and the interior of the U-Boat. Then tell me your plans regarding the storage of the art treasures and the accommodations for our passengers on the submarine."

"Regarding accommodations, I have found a friendly farmer and faithful supporter of the Third Reich, on the town's outskirts who provided me with a decent room in his home. I believe he'll be able to offer you another room, at least until your part of the assignment has been concluded, which should be during the beginning of February. I'll leave for the farm in two hours. I suggest you join me in the Mercedes."

Mertens agreed with this arrangement then asked: "Where can I find Erika?"

"She has left my service without permission," he uttered with a voice of angry contempt. "I could have punished her severely for recalcitrant disobedience, but I have decided to leave matters alone, since she could only cause trouble. I have to think about the future and my special assignment from the *Führer*. Any distraction could only cause severe problems for me. You'll find Erika in the town's hospital, where she works and has a room together with another nurse."

Mertens was greatly relieved to hear this and decided to make his way to town as soon as his business with Kempe was concluded. Thanking him for finding a room, he walked over to Captain Volker's table. He sat alone in a corner with a glass of beer reading the latest newspaper, *"Der Völkische Beobachter"*.

"Do you mind my joining you?" Mertens asked.

"Oh, not at all, please sit down", Volker answered.

"What does the paper say?" Mertens asked. "Tell me about the latest news of the war!"

"This paper is full of lies, and it is necessary to interpret the truth on your own. I just spoke to a friend, who heard from reliable source, that the Russians have started their offensive by entering the German province of *Silesia*. Warsaw has been completely surrounded and Russian troops have taken half of East Prussia, while pressing forward on the way to Danzig and the Baltic Sea. We are lucky to be far away from that area, and I only hope the *Wilhelm Gustloff* will make it safely back to *Lübeck*. I have dispatched a coded message to the *Gustloff* advising the Captain of our encounter with the Russian submarine."

The more the two men talked, the more they took a liking to each other. All Mertens had to do was to voice his negative opinion about Hitler and his Nazis, in order to provoke an angry outburst from the Captain. Volker had instinctively started to trust Mertens, seeing in him an ally with as much contempt for the Nazis as he had. Both concluded the war was lost, and a total defeat would occur within the next few months.

Mertens told Volker about his conversation with Kempe and his request to visit the submarine. Hearing this, the Captain could no longer contain himself. With a grim face, he looked at Mertens.

"As we discussed before, I understand his interest in my U-Boat. These Nazi bastards want to use her to escape with their stolen treasures, avoiding the fate awaiting them after being captured by the Allies. You and I must find a way to prevent this; these murderers of millions must be brought to justice. Don't you agree?"

Mertens sympathized with Volker's opinions, and was elated at this turn of events. He had fortunately found an important ally who shared his own ideals, and who would undoubtedly be extremely valuable in the future.

"May I call you Hilmar?" Peter asked.

"Yes, you can call me by my first name."

"Anyway, I'm Peter!" Then he continued. "These Nazis must be stopped before they can do any more damage to the rest of the world!"

Hilmar Volker's face broke into a broad smile. He enthusiastically extended his hand in a gesture of sincere friendship.

"Peter, we shall keep our eyes and ears open so when the right moment arrives, we can act accordingly; I'll never take any of these Nazi thugs to South America for them to benefit from their stolen treasure. If I didn't have this Nazi Officer on my boat, it would be easy to prevent this planned escape and their undeserved freedom! They are nothing but war criminals!"

*** *** ***

Fighting his way through the ship's gangway, Horst Stolz had a hard time to enter the ship. It was the eve of January 29th. The Gustloff was scheduled to embark very early in the morning. From the eastern horizon a bright reflection of a fiery red sky loomed all over. Continuous rumbling of artillery fire could be heard, adding to the restlessness and anxiety of these refugee masses. Finally, Stolz arrived at his cabin, only to find that he had a roommate, who introduced himself as a Herr Dr. Alfred Kiesler, a scientist and expert on rocket propulsion.

Next morning, Stolz tried to make his way to the open deck and view the departure from Gothenhafen. The ship was packed with five thousand, four hundred passengers, consisting mostly of wounded soldiers, mothers, and children. It was still dark outside, when she removed her lines from the bollards, and the bow thruster slowly maneuvered her away from the pier. Thousands of disappointed refugees were left behind. Stolz returned to his cabin where his roommate was asleep, but started to stir after Stolz had entered. Stolz approached Dr. Kiesler with the thought of asking him for his personal identification papers. There was nothing else to do and he was curious to learn more about this man.

In the semi-darkness he searched for his flashlight so he could read the documents. Suddenly, it hit him like a thunderbolt. He almost dropped his flashlight. No, it couldn't be. The similarity of this situation hit him hard. Yes, of course, Peter Mertens was an American! Now he remembered it clearly. It was during the early morning hours in Berlin and the arrest of the Jewish family, when he found Peter Mertens in his room, and when he checked his American passport.

Stolz panicked. "I've got to do something fast and let Kempe know about my horrendous discovery. The fastest way to notify him would be by wireless from the radio operator's quarters."

Fighting his way through the crowd, Stolz moved upstairs in direction of the Commando Bridge. It took him quite some time before he reached the Captain's quarters. He found him at his post directing the ship toward the Stolpe-Banks in a westerly direction.

"*Herr Kapitän* I urgently have to send a wire to our Eckernförde base!"

The Captain looked irritated at Stolz. "I'll do no such thing. We are under strict orders to keep radio silence, especially since we have been warned that a Russian submarine could be in this area."

Stolz was exasperated and he told the Captain about his discovery. The Captain was also alarmed by the story and promised, as soon as they had reached the proximity of Stettin, that he would order his radioman to transmit this message.

CHAPTER 19

The red brick farmhouse had a large thatched roof. The building faced a small, sandy country road, skirted by tall poplar trees. The farmyard was enclosed by a six-foot high red brick wall, and accessible by a large wooden gate. The wall joined the building at one end, while at the other it touched the dairy cow shed. Further behind laid a hay barn and horse stable. The yard was paved with round cobblestones. Near the bam was the five-foot tall manure pile. A number of chickens, a few ducks and a small flock of white geese inhabited the yard. A grassy knoll with fruit trees sloped from the rear of the property down to a twenty-foot wide stream.

The winter temperature hovered at freezing under a cloudy sky, with the possibility of a light snowfall.

Mertens and Kempe arrived during the early afternoon, and were met by the farmer, a man of about fifty-five, of medium height, with a strong build. Sharp lines around his pale eyes creased his weather-beaten face. His blond hair was cropped in the Prussian military fashion, and on his yellow corduroy jacket lapel he wore a round, enamel swastika pin, indicating he was a proud member of the Nazi party. His brown trousers were tucked inside a pair of brown boots, undoubtedly once part of a full-fledged Brown-Shirt outfit. After a hearty *Heil Hitler* salute, he shook hands with Kempe and Mertens. Then he introduced his middle-aged wife, whose gray-blond hair and very blue eyes revealed her Scandinavian background.

Mertens' room was small but immaculately clean and tidy. The bathroom had to be shared. The farmers wife generously invited the guests to an afternoon coffee of barley brew and a just baked pound cake. Mertens soon learned that two of the couple's sons had been killed on the Russian front, and that a third son was currently serving in the SS-Corps, somewhere in Germany on special assignment.

In order to tend the farm, milk the cows, and handle all the back breaking chores, the farmer employed two Polish prisoners, who had to wear, on their outer garment, the letter 'P' framed by a diamond shaped yellow field. The two young Poles were normally stationed with many other prisoners, inside a nearby barbed wire barrack compound, guarded by SS-soldiers. Every morning at six, the two Poles were brought to the farm for work, and at six in the evening they were returned, under guard, to their prison camp. The farmer was responsible for all their food; however, he was a brutal, mean man, who fed the two workers mostly on boiled potatoes with a smothering of linseed oil. If they didn't work fast enough to suit him, he used a whip on their backs.

"These Poles are lazy bastards," he said with contempt in his voice. "They must be taught what the word 'work' really means. They belong to a subhuman race and must be trained to serve our kind of superior society!"

To Mertens' surprise the farmer's wife objected to this tirade of abuse, insisting the Poles were also human beings to be treated with fairness. Her husband however wouldn't have any of this and he told her in no uncertain terms to shut up.

After enjoying the coffee, Kempe excused himself and drove back to Eckernförde in his Mercedes. Mertens, anxious to see Erika again, asked the farmer if by any chance he had a bicycle he could borrow. The farmer pointed to the back shed and told him to take one of the four, three of which belonged to his sons.

Asking a few locals for directions, he finally found the military hospital. He located Erika in the nurse's quarters, together with two other aides. When she saw Peter, she ran swiftly into his open arms. Erika introduced Peter to her newfound friends who recognizing the situation, excused themselves and left the couple alone. They celebrated their reunion passionately. Afterwards, Peter brought Erika up-to-date about his latest activities. He explained that he still worked for Kempe until the 'Wilhelm Gustloff' had arrived in Lübeck. Also, he still had to supervise the transfer of the new treasures to the submarine base.

The next day was Erika's day off, and they agreed that she should join Peter by walking the four kilometers from the hospital to the farm, a walk of about one hour. Maybe the farmer would lend her one of the other bicycles, and they could both take a short ride through the lovely surrounding countryside.

The next morning, just after Kempe had left the farm, but before Erika arrived, Peter had a visitor. Driving a small Opel Kadett, supplied by the Navy, Captain Hilmar Volker dropped by for a short social call. He explained he was on his way to a conference in Kiel Tannenberg, where Professor Hellmuth Walter had his factory. As it was still early for the meeting, he wanted to see his friend Peter again. Both men made themselves comfortable in the living room, while the farmer and his wife were occupied milking cows. Peter told Hilmar he was expecting his girlfriend, Erika Berger, to arrive at any moment and that he wanted him to meet her.

While they discussed Kempe's request for the upcoming inspection tour through the Captain's U-Boat, they heard a knock. Opening the door, Erika and Peter greeted each other eagerly and then went into the living room. Volker got up from his seat to greet her but both froze in amazement.

"Erika, you here? Where in the world have you been? I'm totally confused! I thought you and your father had to flee the *Gestapo*? Where is your father?"

Peter did not understand what was going on, but his first reaction was to check if the farmer and his wife were still out of sight and hearing. Ensuring that everything was safe he told the Captain to sit down and listen to his explanation: "Hilmar, I'll give you the whole story in a minute, but first tell me, where do you know Erika from?"

'We grew up together in Hamburg. Our parents have been close friends from back in their university days when we were neighbors. I consider Erika family, just like my own sister!"

Peter was dumbfounded and shook his head in disbelief and said: "I first met Erika and her father in Strasbourg, both hidden in an apartment belonging to friends. Learning of their situation, I recognized the danger they were in. After a short time and with help, I succeeded in getting Dr. Rasche across into Switzerland, where he presently is safe."

Peter glanced across at Erika, his voice softened as he continued. "Since Erika and I fell in love we decided to move to Berlin, changing her name to Berger for safety reasons. In Berlin, I had business matters to deal with, and that is where Kempe hired me as an art expert to evaluate a number of art treasures. They all turned out to be looted from conquered European countries."

"I bet I know which apartment in Strasbourg they were hiding in," Volker interrupted.

"It must have been at Mme. Abel, our family's old friend from the Heidelberg University. Well, I am glad the old man is safe, and now you are probably both waiting until the war is over to live happily ever after. But I don't understand why you aren't in the German armed forces."

Peter smiled and pointed to his military *Wehrpass:* "As you see, the Nazis have hired me to handle the transfer of art objects from occupied countries, and therefore I am exempt from military duties. However, I am afraid Kempe is trying to draft me into the SS. Kempe has been my immediate superior and sponsor, but I'll tell you more about that at some other time, because there is much more to my story."

"To exempt you from military duty for such a despicable assignment is incredible. But more power to you if you can get away with it. You realize, of course, that you've become a tool of these criminals. However, I'm glad Erika is in good hands. All we have to do now is to survive this insane war."

They continued talking until it was time for the Captain to excuse himself. He said goodbye and left for the factory conference.

Early one morning, several days later, Mertens, Kempe and the Captain met for the planned U-Boat inspection tour. The Captain, learning of the cargo volume and passenger number, said he could not possibly reduce his crew below three men, plus his First Officer and himself. Consequently, only six passengers could be accommodated. This would leave the forward torpedo area free to store a limited amount of art treasures, and only if all torpedoes were eliminated.

Kempe became very angry hearing this news, demanding that ten passengers must be accommodated. Outraged, he ordered Captain Volker to rethink the matter, and report to him with the final passenger and cargo plan. Kempe stormed off the U-Boat, saying he had to have a reply within twenty-four hours.

Peter and the Captain remained on board, trying to come up with a feasible solution for Kempe. On his way back to his Mercedes, Gerhard Lange, the First Officer of the U-Boat G-17, furtively approached the SS-Officer.

"*Herr Standarten Führer,* I must speak to you in private and confidence."

Kempe was surprised by this approach, and invited the First Officer to join him in his parked car.

"I don't know how to explain this situation to you," Lange began, "but since I've been a long-time supporter of the *Führer* and a party member, I have to express my suspicions. *Herr Standarten Führer,* I must tell you that Captain Volker is not trustworthy. I've overheard him several times making derogatory remarks about the *Führer*, the war and Germany's future. Moreover I heard him say, on various occasions, that the war for Germany is lost and that the *Führer* is mad. Therefore, I believe he should not be entrusted with your most important mission. He also seemed to have struck a friendship with your man Peter Mertens, and maybe you should check on his allegiance to *Führer* and *Vaterland* as well."

Kempe was shocked to hear this outburst.

"These allegations are extremely serious," Kempe said after a moment of silence. "Do you realize you are accusing the bearer of the prestigious *'Ritterkreuz'*, of treason, a hero of Germany? Also my man Mertens has proven himself to me during a most precarious situation. You better watch what you are saying. We all say things on some occasions, which are meant only to let off steam but are not as serious as they sound.

However I am grateful for the information, and I shall keep an eye on Captain Volker and so should you. Thank you again, good bye, and Heil Hitler!"

With these words, Kempe dismissed the First Officer and drove off to make arrangements for the upcoming arrival of party officials slated to leave on the submarines for South America. He also made a request to the two *Gestapo* agents reporting to him, to keep an eye on Captain Volker, and check his political record.

After midnight at the farm, the farmer and his wife were already in bed; a heavy knocking at the front door awakened everybody. Two SS-men had arrived in a side car motorcycle demanding to speak with *Standarten Führer* Heiner Kempe. Kempe and Mertens met the two men in the living room while the sleepy farmer returned to his quarters. After Kempe had ordered the visitors at ease the Sergeant began his report.

"A terrible thing has happened," he said. "The *Wilhelm Gustloff* has been sunk by a Soviet submarine in the area of the Stolpe-Banks, and most of the passengers have been lost. For more details we have to go immediately to the naval station in Kiel."

Kempe and Mertens were stunned by this news. The messengers were dismissed, and Kempe ordered Mertens to join him as soon as they were dressed. Then they drove to the naval headquarters in Kiel to obtain the most up-to-date information about surviving passengers. Kempe was most concerned regarding the whereabouts of his aide and driver, Stolz. So was Mertens, but for entirely different reasons.

One part of the Kiel submarine bunker included the naval station. Despite the early morning hour, at two a.m., the place was in an uproar. The torpedo boat T-36 had arrived with many of the saved passengers from the Gustloff, some of whom were being interviewed by naval personnel. Early reports indicated an enormous human tragedy had taken place.

The callous attack on the passenger liner had been carried out by the Soviet submarine "S-13", which was eventually joined by two additional Russian submarines. The ship was clearly marked with large "Red Cross!" postings. The Russian situation had prevented any German naval rescue mission to take place, fearing for the safety of their ships and the loss of many more lives. Surprisingly, no German anti submarine vessels had been available. The following story eventually emerged from the surviving passengers and crew.

The Russian submarine launched three torpedoes against the unarmed vessel, sinking her within minutes. The escorting torpedo boat T-36 was too far ahead of the liner, and could not attack the Russians. The Gustloff was loaded with over 5,400 passengers, most of them women, children, and wounded soldiers. None of the German ships could come to the rescue, due to the danger of also being sunk.

From the first reports, only nine hundred passengers could be saved, the rest drowned in the icy cold waters of the Baltic Sea. The ship did not have the full speed capacity of their diesel engines, because the repair work in Gothenhafen had not been fully completed. The Russian army was nearing Danzig too fast; the ship had to leave port, regardless of consequences. The normal zigzag course could not be maintained, making it easy for the Russians to hit their target.

Other reports announced the Russians had conquered all of East Prussia, Warsaw and most of Silesia. A new German defense line was hastily established along the Oder River. High school boys and old men were being recruited, equipped with the so-called *Panzer Faust*, a German version of a Bazooka.

Repeatedly, Kempe asked various survivors if they remembered an *SS-Schaarführer* on the ship and if they knew what had happened to him.

"The SS-man! Yes, I remember!" a young woman said. "He had a baton or whip under his arm. He was a brutal character who shouted at us, that he had to be saved, because he had to fulfill a secret mission and he had to transmit extremely important information. Anyone standing in his way was whipped with the baton. On one occasion he even drew his pistol, threatening that anyone in his way would be shot. But suddenly, the ship listed to one side and started to sink rapidly. Everybody was thrown into the cold water. Fortunately, I was picked up by the arrival of the torpedo boat, which rescued a number of floating passengers. I saw your man adrift trying to get rid of his boots. They must have dragged him down because soon I lost track of the *SS-Schaarführer*, and I don't recall seeing him again among the rescued passengers. I am pretty sure he drowned."

Kempe abruptly turned away from the woman without comment. Cursing furiously he motioned Mertens to return to the car for the trip back to their farm quarters. On the drive home, Kempe broke his long silence.

"Peter, your job here is finished; we'll leave the farm this morning at eight o'clock for the hotel "*Zum Weißen Hirsch*." There I'll have to confirm quarters for a group of party members, due to arrive in the afternoon, and with whom I have some important business to attend. As promised, I'll pay you your 3,000 *Reichsmark*. Also as suggested by Adolf Eichmann, I'll take the necessary steps to have you inducted into the SS-Corps. You'll take Horst Stolz's place as my aide and driver. That's not a request; it's an order, taking priority over anything else you might have planned. I assume, though it's evident looking at you, that you are of Aryan blood and don't have any Jewish ancestors."

Peter was horrified by these rapid developments, but knew the war had reached a stage where checking his background was just about impossible. He also realized that this new situation would give him a chance to stay close, hopefully preventing Kempe and his cohorts from escaping to South America.

"I guess this will mean that I can't call you by your first name any more, but will have to address you as "*Herr Standarten Führer*." But if you believe I can be of use, I'd be honored to comply with your request."

"In private you can call me Heiner; however when in public, we must adhere strictly to protocol. Tomorrow we will find the proper uniform for you, together with the necessary arms, and whatever is required."

Forced to deal with this new situation, followed by having to wear the hated SS uniform, all this made Peter extremely uncomfortable. But he saw no other possibility or alternative to the now swiftly-moving events.

As planned, Peter and Kempe left the farm at about eight o'clock for the hotel "*Zum Weißen Hirsch*" (White Hart), where twenty rooms had been reserved, double occupancy for the expected group of party officials.

While they sat in the dining room, the hotel owner approached their table, accompanied by an elderly man who introduced himself as Dr. Franz Weber, Director of the *EWM Werke* factory Lauterberg in the Harz Mountains. He wanted to talk urgently to *SS-Standarten Führer* Heiner Kempe about a very confidential piece of information. After the owner left the dining room, Kempe introduced Peter as *Schaarführer* Mertens, his personal aide, and then invited the visitor to sit at their isolated corner table.

Kempe remembered Officer Lange's warning about Captain Volker and Peter; therefore, he knew he shouldn't trust anybody. He ordered Peter to leave and wait for him in the Mercedes.

Dr. Weber was a thin, pale man of about sixty, wearing a pair of gold rimmed bifocal glasses, and a gray suit with vest. He was obviously in pain, and had great difficulties sitting comfortably. Several times he cleared his throat, his eyes darting everywhere instead of looking into Kempe's face.

"*Herr Standarten Führer*," he began. "Four tanker trucks, each containing thirteen tons of H_2O_2 fuel have arrived at the Eckernförde submarine station.This will be barely enough for one of the four U-Boats.

What is even worse is that the factory will be unable to furnish additional H_2O_2 for a considerable time, because we have lost the promised electric power supply for the electrolysis process. As a result of the continuous bombing raids, our power lines have been destroyed and the electric generating plant has been heavily damaged. We are working feverishly to have the electric supply reconnected, putting the plant back into production. However, it will take at least another month before we can expect to produce enough fuel for one more submarine."

Kempe was furious at the news and replied: "*Mein Gott, Mann*, do you know what you are saying? This is a typical case of sabotage. You know you can be executed for this negligence!"

"*Herr Standarten Führer*, please don't threaten me, I am a staunch supporter of Adolf Hitler, and an old party member. I wear the golden party insignia and would happily give my life for the *Führer* if it would help to produce the necessary H_2O_2 fuel. Unfortunately, the war has turned against our nation, and we are all anxiously waiting for miracle weapons to be used to reverse the war in our favor."

Kempe's rage eventually abated, but he began feeling a terrible frustration at being burdened with all these serious problems. He decided to call a conference with Captain Volker. This time he would insist that First Officer Gerhard Lange be part of the meeting, just in case the Captain should become a political problem. On the spur of the moment Kempe went to the Mercedes and ordered Peter to drive immediately to the submarine base. Peter was told to advise the Captain and his First Officer of an urgent meeting, at nine the following morning.

After Peter drove off, Dr. Weber asked Kempe if he could obtain some pain-killing tablets. In one of the recent bomb raids at the factory, the sudden air pressure of a bomb explosion had thrown him against a wall. His spine was badly hurt, causing him constant pain. Kempe suggested he should go to the nearby hospital, where they undoubtedly would be able to help him. Before they parted, they agreed to meet again at the hotel the next morning.

CHAPTER 20

The hospital was a couple of blocks away from the hotel, and Dr. Weber had no problem finding the building. It was busy and crowded, full of wounded arriving by ship from the east; even the hallways were filled with beds and patients. Eventually, Dr. Weber found and spoke to a nurse, asking her if someone could help him to obtain some pain-killing pills. The young woman showed him to a corner where he could sit and wait, promising to send a doctor to him.

He had to wait about twenty minutes, which gave him time to observe the activities around him. Suddenly, he couldn't believe his eyes. At the far end of the corridor, he was shocked to see Erika Rasche, the daughter of his former colleague from the *Kaiser Wilhelm Institut*.

"How is this possible?" he asked himself. "The Rasches are traitors to the Fatherland and are wanted by the *Gestapo*. Any member of the family should be immediately apprehended!"

At that moment one of the young doctors approached Dr. Weber asking: "Can I help you?"

Completely confused, Weber asked, "Doctor, who is that young nurse at the end of the hall?"

The doctor turned in her direction. "She is Erika Berger, one of our finest nurses!" he said proudly.

Dr. Weber swallowed hard, calmed himself, and explained his ailment, asking the doctor for some pain-killing pills. The physician put his hand on Dr. Weber's shoulder assuring him he would provide some pills if he would wait just a minute for a nurse to bring them. To make sure Erika didn't recognize him, he picked up a newspaper from the table next to him and hid his face.

After a nurse gave him a small container of pills, he left the hospital.

His pains forgotten, all Weber could think of was to get hold of *Standarten Führer* Kempe and have him investigate Erika Rasche, alias Berger.

Arriving at the hotel he saw that Kempe was still talking to two civilians.

Considerably agitated, he went to Kempe and urgently requested a private word with him.

"These two gentlemen are close associates of mine so you can speak freely; please take a seat and tell us what is on your mind," Kempe said, glancing at his two companions.

Dr. Weber, sat down at the table and related his discovery to the three listeners, demanding that someone immediately look into this matter. At first Kempe tried to convince Weber he must have been mistaken, that the nurse he saw merely resembled Erika Rasche. But the fact that the name Erika was used for both names made him finally go along with Weber's request.

"Dr. Weber, you came to the right person. Let me introduce you to my friends, Heinz Hermann and Paul Amtmann, both members of the *Gestapo*. I will send them immediately to the hospital where they will pick up this Erika, whatever her name is, and we shall see what we shall see."

Kempe was suddenly in his element. Hunting and finding people, scrutinizing their backgrounds was his greatest pleasure; especially when the arrested person started to tremble with fear. In the case of Erika, he couldn't wait to get even with her. He ordered the two *Gestapo* men to find Erika, and take her to the local police station, where he would conduct his interrogation.

Less than an hour later Erika was brought to the police station, where Kempe and Dr. Weber were already waiting inside the interrogation room. The two *Gestapo* agents carried her forward, supported on both sides, and pushed her into the room. The daylight from the window blinded her and Dr. Weber, without getting out of his seat said:

"*Guten Tag Fräulein Erika Rasche, wie geht es Ihnen?*"

Erika immediately recognized Dr. Weber as her father's colleague, but she was unable to utter any reply. Her face became pale and her body began to shake.

Then Kempe said with triumphant glee: "I always thought there was something not quite right about you, and now I have proof, you half Jewish bitch! I know all about your Jewish mother. As a matter of fact, we had her done away with to keep your undeserving father at the Institute. Tell us, where is your traitor father?"

Erika was horrified by the news but kept silent. She didn't know what to say. Then Kempe crashed his fist onto the table and yelled at her in a savage fury. "Where is your swine of a father?"

Completely demoralized, she stuttered: "He is in Switzerland!"

"How in the hell did he get there?"

"I don't know, I wasn't with him, he left me while we lived in Strasbourg."

"Tell me, did Peter Mertens have anything to do with all of this? You met him in Strasbourg, didn't you? What is his part in this treasonous plot?"

Erika became frantic and cried out: "Peter has nothing whatsoever to do with any of this."

Kempe instinctively knew Mertens was somehow involved, and immediately ordered the two *Gestapo* agents to go to the submarine base, arrest Mertens, and deliver him straight to the police station.

"Look Erika, you might just as well come clean and tell me all about Mertens, including your spying activities, who you are working for, and what you are planning against Germany, your own country?"

Erika had a hard time to control herself, but she was determined to endure this humiliation and she kept her mouth shut. Kempe, realizing he would get no more information for the moment, asked the Police-Chief that she be placed in one of the available cells until further notice. Kempe thought for a while, what his next action should be, and decided to play Peter and Erika against each other, in order to get some answers; failing that, he could always resort to torture. Too bad Horst Stolz was no longer around; he could have extracted information very quickly from this pair.

Due to the imminent arrival of party officials, Kempe left instructions to place Mertens, for the time being, in one of the other cells, away from Erika. He also instructed the head of the police station to keep the two prisoners isolated until he returned next day to continue his interrogation. Then he told Dr. Weber to accompany him back to the hotel.

By late afternoon most of the officials chosen for the submarine escape had arrived; however, some of them failed to make it, having become victims of the area air raids. In fact, at the final count only thirty-five men had arrived.

Kempe remembered he was to meet with Captain Volker, together with his First

Officer, at nine o'clock, the next morning. At that time, the final decision regarding the availability of submarines and the number of passengers had to be made.

Now the exciting challenge of interrogating traitors forced him to consider new priorities. For the time being, he planned to have his party comrades settle down in the hotel, postponing all business until the next day. Moreover, it would be beneficial, he thought, to let the prisoners stew in solitude, while he handled the passenger problem and space requirements for the submarines.

Several hours later the two *Gestapo* agents arrived back at the hotel advising Kempe that Mertens was now behind bars in one of the station's jail cells. They also returned the borrowed Mercedes, which was now parked out front.

Kempe inquired:

"Did Mertens give you any trouble?"

"Not at all," Paul Amtmann replied. Mertens was in the cafeteria in deep discussion with the submarine Captain when we approached, and before he knew what happened, we had handcuffs on him. Of course he protested, but we had our orders and we know our business."

"Excellent!" Kempe answered gleefully.

The next morning at the hotel, Kempe met with the two U-Boat officers and Dr. Weber in a separate room, away from the inquisitive party officials. Captain Volker was very upset about Peter's arrest, and wanted to know what had happened. Kempe explained that Peter was suspected of working for the Allies, but in a couple of days he would have more information; especially after some intensive interrogation had been concluded. Volker was alarmed by this piece of information, but felt instantly it was better to wait cautiously for Kempe's findings.

Disregarding further discussions on this subject, Kempe asked Volker's opinion about the embarrassing H_2O_2 fuel shortage. Volker was surprised to hear this news, but was secretly pleased that the Nazi criminals had run into this problem. Concealing his real feelings, he explained that the submarine had a submerged running range of only some 125 nautical miles with a full H_2O_2 storage capacity of 55 tons. Reducing and sharing this fuel with other U-Boats would render the escape features totally ineffective. Indeed, just to get away from the European coast in a submerged state would probably alone require a minimum of 125 nautical miles. He visualized diesel powered surface travel north around the top of Denmark, into the North Sea. The trip through the narrows of the English Channel would be done in a submerged state, followed by a southwest surface course into the open Atlantic and down to South America. Of course at the first sign of enemy surface presence, it would become necessary to submerge and use the H_2O_2 power again.

Kempe became frantic. He had to report this negative news to Eichmann but couldn't reach Berlin; all connections seemed to be cut. He had only one goal in mind, and that was to get the hell out of Germany before he and his party comrades were captured by the Allied Forces. The final decision on the evacuation now rested on his shoulders.

Captain Volker was willing to increase the number of passengers from six to seven as a final concession; on hearing this, Kempe made up his mind. He decided to inform the Admiral in person of these latest developments, and ask his advice on what to do with the three, unfueled, now obsolete U-Boats. Kempe knew the Admiral was presently situated at the old naval school in Mürvik, a village near Flensburg. He decided to drive immediately the 65 kilometers to him, accompanied by Dr. Weber as a witness.

The naval station was an old red brick building, with a large, sloping, tile covered roof. The Admiral had moved here from the Berlin Tirpitz Ufer headquarters, and made this his new command post, from where he intended to conduct the rest of the war. After passing two armed naval guards, the reception desk officer received permission by telephone to have Kempe and Dr. Weber enter the Admiral's office.

On entering the room and after the obligatory Nazi salute, the Admiral asked them to be seated. Kempe explained the dilemma of the H_2O_2 fuel supply at length. Dr. Weber supported all of Kempe's statements. After learning this, the Admiral was silent for several minutes, and then started to speak:

"Gentlemen, we are now in the middle of February, 1945. The Allied armies have penetrated Germany. The Soviets are at our Oder River defense line, where we hope to contain them until our -miracle weapons can be used. In the west, the Americans and British have crossed the Rhine River and are on the way to Bremen and Hamburg. It won't take long before they reach our area. With the disastrous shortage of H_2O_2, we are limited to only one of the four high-speed prototype submarines. I can tell you that fortunately, we have just completed two revolutionary Electric U-Boats, which could as well be used for the escape to South America. They are capable of traveling submerged at 17 knots under battery power only and for a number of days. Their large size built could easily accommodate the rest of the party officials chosen by our *Führer*, for the trip to South America.

"Accordingly, I order you to destroy the three unusable Walter U-Boats by scuttling them 50 kilometers west of this coastal point and south of the Danish isle of Aero. Send all remaining surplus passengers from the U-Boat G-17, to the Bloom and Voss shipyard in Hamburg, immediately. Two Electric U-Boats have already been launched, tested and are now available for boarding and the trip to South America. I'll advise Blohm and Voss immediately of the impending arrival of the personnel. I'll also tell the *Führer.*"

"Herr Admiral," Kempe said. "I would like to request the replacement of Captain Volker with his First Officer Gerhard Lange for the command of my U-Boat. Lange is a good party member and a staunch admirer of our great Führer. I would be more comfortable with him in command on this trip."

The Admiral was not surprised to hear this request; understanding fully that Kempe would want to have a totally party-committed comrade at his side. The Admiral also had heard rumors of Volker's contempt for Hitler's Nazi party officials.

"Your request is granted; as I wanted to promote Lange to command his own boat anyway. Now I need someone who can handle properly the scuttling of the remaining U-Boats, and I can't think of a person better suited for this task than Captain Volker. I'll take steps immediately to arrange for the change in commands. Now I wish you good luck, *Sieg Heil* and *Heil Hitler.*"

During the early afternoon, the two men returned from Mürvik to the hotel in Eckernförde where Kempe called an immediate meeting of his Nazi comrades delegated for the escape from Germany. He explained the situation, repeating the information he had received from the Admiral. He then picked six members of the group, who were familiar to him from his active days in the SS-Corps, and ordered them to accompany him on the escape trip aboard the Walter U-Boat G-17. The rest of the group was instructed to make their way to the Blohm and Voss shipyard in Hamburg by taking an especially reserved bus awaiting them at the hotel.

The barge Captain Günter Müller was still in Eckernförde, and was instructed to load three quarters of the looted treasure items from the U-Boat bunker onto his barge, then transport them swiftly to Blohm and Voss. Kempe was relieved to have solved the escape problem so smoothly. Now he had time to handle the other problem, namely, Mertens and Erika, for whom he had planned a most sadistic, terrifying punishment and execution.

CHAPTER 21

Peter, sitting in his cramped jail cell, couldn't imagine where the leak had occurred or what discovery had been made to cause his sudden arrest. Nobody had yet told him why he had been picked up. He had now spent a night in this police cell without knowing what lay ahead or what he was charged with. Suddenly, during the morning he was taken from his cell, hand-cuffed by two policemen and pushed to the station's interrogation room where Kempe with two Gestapo agents was already waiting. With their backs against the wall, they were sitting behind a table on which bright glaring lamps stood ready to penetrate Mertens' eyes.

He had to stand in the center of the room, a table and chair between him and the interrogators. Nobody spoke; finally Mertens asked what this ridiculous arrest was all about. Hearing the question, Kempe leaped from his chair and exploded with vicious, angry accusations:

"You have made a fool out of me! I know that you are a spy working for the Soviet Union, trying to sabotage all my efforts. Fortunately, the Almighty is protecting the Führer and our party; he is on our side and has arranged for us to discover all of your subversive schemes!"

"Do you remember Dr. Weber? Well he is the one who discovered your sweetheart Erika Berger alias Rasche at the hospital, fortunately without her seeing him. Dr. Weber worked with Dr. Rasche at the Kaiser Wilhelm Institut and knows all of the treasonous Rasche family members. I also know Rasche is in Switzerland; because Erika has confessed to everything and accused you of being a spy."

Mertens was devastated by this news; especially hearing that Erika was in Kempe's hands. He couldn't believe she had betrayed him and assumed Kempe was bluffing to have him confess.

"I don't believe a word of this crap, or that you have a shred of evidence against me. I am a good German. All you are interested in is getting even with

Erika and me; because she rejected your clumsy attempts to rape her."

Kempe exploded into an uncontrollable, almost insane rage. He jumped up again from his chair, rushed over to the defenseless Mertens and hit him with both fists on the face until blood gushed from his nose, mouth and lips. Peter understanding the typical egomaniac mentality of this fanatic Nazi, acted as if the blows had been enormously powerful and purposely let him-self fall to the floor. Kempe felt a great deal of satisfaction, superiority and physical strength. The two Gestapo thugs approved of Kempe's attack with grins over their pig like faces.

The two policemen pulled Mertens roughly to his feet, seating him on a chair. Kempe returned to his seat, trying to control himself so he could continue his interrogation.

"I suggest you tell me about your espionage assignment and who you are working for, otherwise I'll torture Erika in front of your eyes until I have obtained every detail."

Mertens knew he was caught in an impossible situation, for he would never permit Erika being tortured. His senses racing, he started to create what he hoped would be a convincing story.

"I know you are capable of any type atrocity!" He began. "So I have no choice but to confess everything. The information will not benefit you in the slightest, since Germany has lost this damn war and within weeks you'll have to answer to our war crime tribunal, which will bring you to the end of a rope. Yes, I am an agent, but not for the Russians. I am an American Captain of the Ranger Battalion, specially assigned to keep an eye on you Kempe, and report your war crime atrocities to my people and the world. The escape from Paris in the Sherman tank was only a ploy to lull you into trusting me. I am happy to say my mission has been successful, and all information regarding you and your Nazi criminal friends, including your escape plan, is already in the hands of the Allies."

Kempe swallowed hard at this revelation for he knew Mertens was right; the war was lost and any information he had or could extract from him had no value, except for personal satisfaction. Therefore he decided to wreak his own kind of cold-blooded revenge on Peter and Erika.

"You and your whore have no future!" He announced with a smirk of sadistic gratification. "I shall slowly drown you both in one of the now obsolete submarines to be scuttled in the Baltic Sea. You two will suffer like rats caught in a trap. We'll not touch you, or torture you beforehand, so you'll be fully aware and conscious of your misery, and suffer every moment of your last lung searing, gasping breaths...."

With this, Kempe dismissed Mertens to his cell, instructing one of his agents to have Captain Volker meet him at the hotel during the afternoon for further instructions.

As requested Volker showed up, meeting Kempe and the two Gestapo agents inside a separate hotel room. Kempe opened the conversation by telling of Erika and Peter's arrest and spy activities. Volker was deeply shaken; it took all his power not to show his dismay and concern for Erika. He was not surprised to learn that Peter was an American spy, nor did this knowledge destroy his friendly feelings toward him.

Kempe asked Volker if he had heard anything from the Admiral.

"Yes, last night I was ordered to the Admiral's headquarters in Mürvik. He instructed me to take a new assignment, and relinquish the command of U-boat G-17 to First Officer Gerhard Lange, who has been promoted to Captain. My new assignment will be to move three Walter H2O2 submarines into the Baltic Sea, south of the Danish isle of Ärö, where I am to scuttle and sink these vessels!"

Kempe was gratified to learn that the Admiral had followed his wishes.

"I'll contact the Admiral to have these instructions verified!" He told Volker. "But for your advance information, I want you to scuttle only two of these U-boats. The third vessel will be needed for a special purpose; I will decide the timing for its eventual scuttling at Ärö personally. I would like to use the boat for drowning two special rats. Therefore, I want you to advise me immediately, when you have scuttled the other submarines. Do you understand?"

"Jawohl Herr Standarten Führer!" Volker replied.

After Volker left the hotel he thought about Kempe's instructions, and on reflection it seemed evident, that the Admiral wanted to remove him from his command of the G-17. Volker fully understood that Kempe would be more comfortable with a Nazi confidant during his trip to South America. However, what really bothered him was the instruction for the third submarine. Fortunately, Kempe had, undoubtedly, revealed his intentions to drown Peter and Erika inside the u-boat.

Suddenly he remembered the last conversation with Dr. Rasche, when they met at the submarine bunker cafeteria, just before Dr. Rasche's escape to Strasbourg.

"You will assuredly know what to do, when the right moment presents itself!" The scientist had said.

There was no doubt in Volker's mind that the right moment had indeed arrived. There had to be a way to stop Kempe. Volker concentrated all of his thoughts in that direction. He had to make preparations without causing suspicion, for only he could handle the task. Exact timing was an absolute necessity for success. Getting a message to Peter advising him of his escape plan on one hand; while organizing and executing the preparations would be a major difficulty, on the other hand.

By the beginning of March 1945, Volker had accomplished the scuttling of two U-boats off Ärö and advised Kempe accordingly. Kempe in return set the exact date when Volker should bring the last U-boat to its scuttling location, aided by a crew of six men, having the two prisoners ready for action plus two Gestapo agents.

This arrangement gave Volker enough time to make preparations, although he still had the problem of getting his instructions, for a successful escape, into Peter's hands. They were written on a piece of folded waterproof paper. With a wax pencil, he had printed clear directions on how to proceed in escaping from the sinking U-boat.

Several days later, early in the morning, Mertens was led from his cell; his hands tied with thick rope, because the police didn't want to lose their valuable handcuffs. Shortly afterward Erika, whom he hadn't seen during their imprisonment, joined him. He was deeply concerned how she would hold up under these terrible conditions; she looked as if she had suffered and been crying a lot, which certainly was understandable.

A police van drove them to the submarine base from where they walked to the U-boat mooring. The crew and their Captain were waiting on board. After Volker gave instructions to enter, they all embarked through the forward deck hatchway into the boat's main body by climbing down a steel ladder. Kempe and his agents brought the prisoners to the aft section of the boat where they roped their hands to one of the exposed horizontal pipes, preventing any further movement. At this point Kempe gave orders to Volker to cast off and move toward the U-boat's destination.

The powerful vibrating clatter of the diesel engines indicated that the vessel was ready to leave. Shortly after departing the mooring, a sea going tug followed in their wake to return the submarine crew plus its passengers to Eckernförde.

Peter was unaware that it would take about five hours to reach the target area south of Ärö. Meanwhile he desperately tried to think of how he could free Erika and himself. The only possibility immediately in sight was an adjacent pipe flange with protruding studs and nuts; possibly they could serve as an abrasive surface over which he could rub the ropes until they broke. If he only could move! He tried to encourage Erika to be brave, in a voice loud enough to carry above the sound of the engines.

Captain Volker having completed piloting his vessel through the harbor area from the conning tower, finally found time to go below and survey the situation. He spoke to his crew giving them orders regarding their duties in scuttling the submarine. Together with one crewmember, he walked aft to the engine room to inspect the planned opening of a deck hatchway. In doing this they had to pass the prisoners. Kempe's watchful eyes followed them closely.

After the inspection and on the way back to the control room Volker walked behind the sailor using him as a shield from Kempe's constantly inquisitive eyes. When they passed the prisoners again, Volker slipped his escape instructions into Peter's outer jacket pocket, together with a small but very sharp knife, and fortunately without being observed. Peter noticed something was put into his pocket and wondered what he would eventually find.

Kempe being most suspicious walked over to the prisoners checking to see if everything was normal, thoroughly inspecting the tightness of the ropes. Satisfied that everything was in order he returned to the control room area.

Several hours later having reached the target area the Captain stopped the diesels and ordered everyone to leave the vessel so he could commence the scuttling operation. Kempe insisted on staying aboard, remaining with the Captain until the last possible moment.

The crew went to work by opening all hatchways on the fore and aft decks including the conning tower. Finally, everybody left the boat except Volker, Kempe and two crewmembers who now operated the opening of the ballast dive tanks.

The sound of in-rushing water broke the silence. The two sailors hurriedly disembarked through the conning tower. Volker ordered Kempe to leave, which he did reluctantly and only after he had once more assured himself that the two victims were still solidly tied down. Then the Captain indicated that he'll be the last one to leave the vessel.

Volker having previously sunk the other two U-boats was fully familiar with the timing of the sinking process, but decided to hurry Kempe by shouting at him to leave more quickly. As Kempe climbed over the outside of the conning tower, the Captain raced to the aft section of the boat, took the knife from Peter's pocket and cut the ropes. Within seconds he was back in the conning tower, just as Kempe poked his head down and angrily called for the Captain to show himself. Volker needing an excuse for his delay replied he still had to shut down a few instruments in the tower. Then he joined the rest of the disembarking men, stepping on to the waiting tugboat. Now they moved away from the submarine to a safe distance. Kempe insisted to watch the sinking U-boat including the plummeting of her doomed prisoners.

Volker had instructed his men to open the ballast tank valves only in a half-open position. He wanted to be sure that the sinking process would be slowed down. This delaying tactic gave Peter time to cut Erika's ropes, then remove the note from his pocket and look at the message. The visibility was aided by the illumination from the emergency lights.

Following the instructions precisely he immediately closed the water tight compartment door to the front torpedo room, then did the same aft, sealing off the engine room. This would gain Peter precious time until the very top of the conning tower edge would sink below the water's surface. This arrangement would help to delay the sinking time; until the centrally located control room and Peter's position would also be flooded.

The next direction told him to head for the Captains quarter, where he would find two wetsuits with diving masks, lead belts and air tanks under the bunk. Erika removed her white nurse's skirt and Peter helped her to struggle into the diving wet suit of porous rubber skin. He strapped the floating kapok vest and air tank over her shoulders and placed the gasmask- type breathing apparatus over her face. Then he attached the lead belt around her waist.

Peter instructed her that, at his signal, she should carefully open the air valve attached to the flexible hose, coming from the supply tank and leading into the mask. In the meantime, she should put her fingers between the mask and her face for normal breathing.

As Peter and Erika made their preparations, the incoming water rushed in from the two deck hatchways, flooding the forward and aft sections of the boat. Together with the ballast tanks the boat sank sufficiently to start an overflow of seawater through the conning tower top rim. Peter knew he had to hurry and slip into his own diving gear.

Eventually the water, in a roaring torrent, gushed into the control compartment, the level rising rapidly around the two intended victims. Within seconds they had to open the diving mask air supply, as water started to rise above their heads. Submerged in seawater, groping for each other, Peter guided Erika to the conning tower exit where the flow of incoming water had stopped, because the control area was now fully flooded. He moved Erika carefully toward the tower, through the opening, and away from the U-boat, which soon silently sank toward the bottom of the sea without them.

Above, daylight penetrated the ocean's surface, while below the interior emergency lighting of the U-boat still shone upward through the open conning tower. Holding onto Erika, he had to prevent her from acting on her natural instinct, wanting to surface as fast as possible. Fighting this urge, but aided by the lead belts, they had to adjust and counteract the diving gear's slight buoyancy; they moved arms and legs against this updraft, remaining under water. Peter still able to see the hull of the tug from beneath his position wanted to be certain that Kempe and his associates were well out of the area before attempting to emerge on the water's surface.

Kempe watched the sinking of the submarine with great satisfaction and decided to demur to be absolutely sure neither victim would float to the surface. After ten minutes had elapsed he gave orders to the tug Captain to return to Eckernförde.

From Peter's underwater position he could clearly hear the sound of the tugboat's engines and feel the rotation currents from her propellers. After the noise had abated they released their weighted belts and both floated leisurely to the water's surface. The thin layer of water between the inner surface of the wetsuit and their skin kept them from freezing to death in the cold water. It provided an insulating layer of natural body heat and thermal protection.

CHAPTER 22

Peter and Erika floated on the choppy surface of the Baltic Sea, holding on to each other so as not to drift apart. They took off their air supply tanks and face masks to let them float away. Becoming waterlogged, they discarded the black kapok filled life jackets around their necks which had lost most of their buoyancy. As they rose and fell with the sea swell, they barely saw the naval tug with Kempe aboard disappear from their view.

They were alone in the unfriendly sea. The water was cold, but the rubber skin wetsuits did their job protecting the swimmers from initial hypothermia and a cold, shivering death. Looking north they saw the low contours of land of the Danish isle Ärö.

Peter shouted to Erika: "Keep swimming toward the shoreline".

He estimated the distance to be a minimum of six kilometers, which, according to his calculations, would take at least three hours to reach land, provided no currents slowed their efforts. Fortunately, Erika was a good swimmer. The only main problem was the very cold water.

At about two in the afternoon, the weather was gray and overcast as they drifted and swam toward the less turbulent waters of the lee shores of Ärö. The sea had calmed down somewhat, and the wind had dropped to a light breeze. But in spite of their steady swimming, the cold had penetrated their bodies, making Peter more and more concerned that they might not reach land. Erika seemed to fare better than he did, even though she too was very cold, but Peter knew that women could take low temperatures better and longer than men could.

After a little more than half an hour, Peter thought he could make out a small vessel moving ahead of them, growing in size as it got closer. It looked like a fishing schooner heading in their direction. On its stern flapped a red flag, which made Peter extremely worried, fearing it might be a German fisherman, who would undoubtedly turn them in to the authorities. However, they were freezing to death, so in spite of his misgivings, he raised his body as far up as he could, out of the water, waving his arms and hoping the vessel's crew would see them. As the boat came closer the flag waving from her stern became quite clear; it was the *Danebrog,* the red flag with the white cross of Denmark.

The skipper had seen something floating in the sea and came to investigate. The diesel-driven vessel came toward them against the wind, so as not to run them down. One of the fishermen threw a rope with a bowline sling at the end, large enough to be slipped around the body under the armpits. Two strong fishermen then pulled the frozen and exhausted couple, one by one, aboard.

"*Hvor kommer du fra*?" (Where do you come from?) Was the question the Danes asked, but both Erika and Peter were unable to speak. They could not speak for their shivering and their teeth chattered continuously. The evaporating water from their bodies and the surrounding air movement increased the excruciating effect of the cold, which stabbed their skins, as if with blades of ice. They were guided swiftly down below deck, and handed two warm blankets and hot drinks. The fishermen left the two lucky survivors alone, giving them a chance to get out of their wet suits and wrap themselves inside the warm blankets.

Speaking German, and now able to talk coherently, Peter called the fishermen, asking one of them to come below. Snuggled against each other, still shivering in the corner of a bunk, Peter and Erika were trying to get warm. The Danish skipper showed much compassion and sympathy for their ordeal; however, his German was poor and intermixed with Danish. This prompted Peter to try English, and to his total surprise the skipper spoke it fluently, so Peter told him what had happened. The skipper shook his head in amazement.

''We had noticed already a couple of days ago that the Germans were scuttling U-Boats off our coast," the skipper said. "And that was the reason we came in this direction, to check if any debris could be recovered or if some valuables could be salvaged."

While speaking with the Danish skipper, Peter had the opportunity to examine his rescuer. He saw a man of about sixty, his weather-beaten face dominated by sky blue, friendly eyes. A healthy growth of gray hair was covered by a dark blue sailors cap. Peter asked the skipper where he had learned his excellent American-English, to which the man replied that he had been living for fifteen years on Long Island, from where he operated a fishing boat out of Montauk Point. As he got older he became homesick for Denmark, so he and his wife Dagmar decided to return to family and friends on Ärö. He introduced himself as Roland Sörensen, explaining that the two sailors on board, were his sons, Helge and Knut, both born and raised on Long Island.

At this point Peter decided to tell Sörensen who he really was, and that Erika was part of his mission. He said it was vital for him to return immediately to Eckernförde in order to complete his assignment and be available with up-to date information when the Allied Forces arrived.

Sörensen was intrigued by Peter's story, offering any help he could provide in supporting his remaining mission. Peter thanked him for their rescue and his offer of staunch cooperation, replying that he had to think about his next move, before involving him and taking advantage of his brave help. In the meantime, his sons on deck were taking the boat back toward Ärö and their homeport of Marstal.

Sörensen suggested Peter and Erika stay below, until the harbor area was clear of the small detachment of German soldiers, still guarding their community. In the meantime Sörensen promised, after docking, to go ashore and return with some clothing.

Later at the fisherman's home and with typical Danish hospitality, the two sons moved into one bedroom, leaving the other to Peter and Erika. After a good night's rest, they all met for breakfast in the dining room, around a polished wooden table and six chairs made of local pine. A glass case on the wall contained a variety of Danish Royal porcelain and some *Hummel* figurines.

After one of the Sörensen's sons had checked outside the house to make sure no German soldiers were in the area, the conversation continued in English. Erika's was passable, but carried a strong Berlin-German accent.

After Peter mentioned again that he had to find a way to get back to Eckernförde, Sörensen came up with an idea of how to achieve this feat.

"I have a German cousin who lives in Flensburg, just on the other side of the Danish border," he said. "He is about my age. His name is Holger Pedersen, who really is a Dane through and through. He hates the Nazis with a passion. Many times during my fishing trips, I run my boat to Flensburg to see him. The German authorities have seen me many times before and know my schooner, 'Hilda'. I'm always welcome, especially if I bring along the results of a successful catch. I could make a trip to Flensburg with you aboard. From there to Eckernförde is not a big distance, no more than sixty kilometers."

Peter was more than pleased with the plan so they agreed to set out for Flensburg the next day, a trip of maybe five hours with good weather and reliable diesel.

Early the next morning at five o'clock when it was still dark, Sörensen, his sons and Peter stepped again aboard the life saving "Hilda", a motorized sailboat, rigged as a two-mast gaff schooner. Erika stayed behind, imploring Peter to be careful and hurry back. The cool wind was about eight knots from the southeast, and the sky displayed a diamond-studded dome of brilliant stars.

The boat's diesel moved them out of the harbor. As the contours of the island slowly disappeared into the darkness, the skipper turned the boat into the wind giving orders to stop the engine and hoist the main and foremast sails. Completing this task, the vessel turned down wind, heading toward the west. Finally they also hoisted the jib and staysails. The main and foremast sails were both lifted, via block and tackle operated halyards, easing the load of the heavy gaff attachments.

There was silence after the diesel was shut down. The soothing wave action of sea against boat was the only hissing sound to be heard. The boat ran winged out before the wind at a speed of about seven knots or almost thirteen kilometers per hour. At this pace, they would be in Flensburg before noon.

Peter stayed below as the 'Hilda' entered Flensburg harbor with reefed sails and under diesel power until they finally moored at the city pier. In the late afternoon, when it got dark, Captain Sörensen invited Peter to join him by visiting his cousin's home, leaving the two sons behind on the boat.

Holger Pedersen lived in an old house near the harbor area, and after they were greeted and seated in the living room, Pedersen introduced his Danish born wife Birgit. Roland Sörensen took this occasion to tell his relatives the truth about Peter, and they were delighted with the possibility of helping him to get even with the hated Nazis.

He was provided immediately with a room, where he could rest and plan his next moves. There were two possibilities of reaching Eckernförde, one by using the railroad, the other by riding a bicycle. It was decided the bicycle was by far the safest way to travel, especially since the Pedersens had one to spare.

Peter knew that returning to Eckernförde would be dangerous, because somebody might recognize him, so he planned to enter the town in the disguise as an old sailor. The Pedersens offered to provide the necessary seaman's gear. Suddenly Birgit Pedersen let out a cry and said she had a brilliant idea.

"Herr Mertens, I have been a professional make-up artist at the Flensburg

City Theater for many years; therefore, I could surely help you alter your facial appearance," she said. Due to Dr. Goebbel's latest decree, all our theaters are closed, but I can return to my dressing rooms and obtain anything needed for your disguise. Let me leave for the theater now, and I'll return shortly with some items to help you and change you into a different person."

In less than an hour Birgit returned from the theater with enough make-up for a theatrical production. In the meantime Captain Sörensen had left again to embark and sail with his sons back to their home on the isle of Ärö. Peter at that point decided to rest until next morning, at which time he would wear his disguise.

The next day Peter dressed in a pair of old brown bell-bottom corduroy pants, a well washed dark blue turtle neck sweater, a faded navy blue short jacket with double breast buttons, and a dark blue sailor cap with visor.

Then Birgit Pedersen started to work on his face, after protecting his upper clothing with toweling. .The first thing she altered was his well-shaped nose, by forming it into a bulbous protuberance using theatrical putty. Then she added a bushy pair of gray eyebrows, changed his temples to a similar color, and finally added a full-size gray beard to his face. Everyone swore even his own mother would not recognize him. The Pedersens gave Peter two hundred *Reichsmark*, along with a few food stamps, so he could survive the next few days. Mounting his bicycle he left these wonderful people with a hopeful farewell: "*Auf Wiedersehen*".

The six-hour bicycle ride to Eckernförde was uneventful and he arrived at the outskirts during the afternoon. Peter knew Captain Volker had rented a room at a boarding house not far from the hotel *Weißer Hirsch*, so he decided to wait for him inside a small nearby park enclosure, which provided a bench to sit on. In the middle of March the weather was overcast with a mild temperature of fifty degrees, making one almost feel that an early spring was approaching.

Finally, at five o'clock, Peter saw Volker arriving on his bicycle. Walking over to the building entrance, he called to him to wait. Volker turned toward Peter wondering who this stranger was and what he wanted from him. When they were two feet away from each other Peter said. "It's me, Peter Mertens in disguise!"

Volker gulped, looked hard at Peters face and tried to find some familiar features, but he was speechless in amazement, wondering how his friend changed his appearance.

"This is incredible; thank heaven you survived, where have you come from?

Who made you look like that? What happened, and where is Erika? Please tell me everything!"

Peter suggested Volker follow him to the park bench, where they could talk without being overheard. He told him what had happened and where Erika was. Volker could only shake his head, saying he had hoped the Danish fishing vessel would show up, just as it had done before, when he scuttled the other two submarines.

Volker told Peter that Kempe was bragging about his brilliant capture of two spies. Kempe had set a date for embarkation and departure of U-Boat G-17 for March 20th, two days from now, to include himself and six other Nazi-bigwigs.

"Something has to be done quickly to stop Kempe from leaving," Peter said. "We must try to screw up the embarkation!"

Volker wholeheartedly agreed. They arranged to meet again next day at the same place, checking to see if either one of them had thought of a way to stop the Nazi escape attempt. Before parting, Peter asked Volker if he could get him a pistol or revolver and some ammunition. Volker said he'd take care of it tomorrow.

On the way into Eckernförde, Peter noticed a huge haystack covered by a black tarpaulin standing in a large meadow. He decided to sleep there for the night. Nobody would notice him there and he could hide the bicycle inside the hay.

CHAPTER 23

Peter awoke early the next morning, having slept fairly well inside the warm haystack. But now standing outside in the cold light of dawn, he knew it was time to return to the place where Kempe was headquartered. It was only a short bicycle ride to the farm, where across from the house, a few birch trees and bushes served as an ideal observation spot.

Parked in front was Kempe's Mercedes. As expected, at eight o'clock he appeared, accompanied by the two *Gestapo* agents. Kempe had apparently decided he needed bodyguards, for whom he obviously had found quarters at the farm. This unexpected situation complicated matters; nevertheless Peter knew he had to stop Kempe.

After the Mercedes drove off, Peter followed on his bicycle but within a short time the car was out of sight. He knew Kempe would go either to his party comrades at the hotel or return to the submarine base, probably the latter, especially since he had set his departure date and time for early morning the next day. Peter opted to continue cycling toward the harbor area in Eckernförde.

Arriving at the waterfront, the number of small sailing boats gave him an idea of how to deal with Kempe. One boat in particular, a simple, one mast gaff rigged sloop, caught his attention, because an elderly man, apparently the owner, was tinkering with it.

Peter approached the man with a cordial, "*Guten Morgen*", which the old timer echoed. Soon they were engaged in a friendly conversation on the pleasures of sailing and the peaceful isolation it can provide. Peter introduced himself as coming from Berlin to flee the onslaught of the Russian hordes, and await the end of the war in this relatively safe haven of Eckernförde. The old man reacted with a cautious nod of understanding, introduced himself as Walter Hellwig. He agreed with Peter's decision to seek safety in the area.

'When do you think the war will be over, and what do you think will happen to Germany?" he asked.

"Well, I believe we will be overrun by Allied forces, and partitioned, or divided into occupation territories for a long time, until Germany can rule itself again. In the meantime, war criminals will be caught and tried. I am happy to say I never belonged to the party."

The man was surprised at Peter's candid opinion, but very suspicious. Maybe he was dealing with a *Gestapo* agent trying to trap him.

'What you are saying makes a lot of sense," he replied. "The main fear I have is, what kind of treatment we will get from our enemies; especially the Russians, who must have an enormous hatred for us. Too much blood has been shed."

Then changing the subject he asked what kind of job or profession Peter had. Peter explained that he was a barge sailor, occasionally involved in shipping art works along the Elbe River, and that he had been hired to safeguard national treasures from the threat of bombs. The man was impressed, and asked what kind of artwork were being shipped; mentioning that he himself had great fondness for French impressionist paintings. He started to name a number of artists and paintings Peter was fully familiar with.

As the conversation continued, the two men realized they had a lot in common with each other and eventually, as Peter had hoped, the man said: "l plan to go out sailing for a couple of hours; would you like to join me?"

Peter agreed eagerly and soon the two men hoisted the main sail and were headed out into open waters. The weather was not too cold.

The man revealed he had never been a party member, and was glad that the war seemed to be coming to an end. He had a son drafted into a FLAK unit, located somewhere on the western front, whom he hadn't heard from in a long time, but hoped would survive the war unscathed. He explained he owned a bookstore in town where most of his business consisted of loaning out used books, as new ones were no longer available.

Peter sensed from his comments, that the man was not at all fond of the Nazis, so he voiced some strong criticism of the murderous brown thugs and their leader Adolf Hitler. He felt that the man, who frequently nodded his head, fully agreed with his comments.

The conversation continued and Peter promised to visit his bookstore, but in the meantime he wanted to ask a favor.

'Would it be possible to go sailing again tomorrow! I would like to bring a friend of mine along, who also loves the water, and sail boats in particular."

"No, I'm sorry but tomorrow I must tend to the store, my wife, who usually looks after the shop, has made other commitments and cannot be there."

Peter produced one of the hundred *Reichsmark* bills and said: "Could I rent the boat for a couple of hours?"

The man rejected the money and looked at Peter. determining whether he should trust him with his boat. After a moment's thought, he agreed to let him take it. Peter insisted he take the money, which he eventually did, pocketing the hundred-mark bill after some protest.

Once they had returned from sailing, the man chained the boat to the pier with a key operated safety lock, then took the key and hid it under a special plank on deck, advising Peter to remember this spot. They separated and Peter promised to show up at his store once he had completed the following day's sailing trip.

Cycling back into the city, Peter went to the hotel *Weißer Hirsch*, where he wolfed down a hearty lunch. While eating, he noticed the obviously six party officials, slated for next day's escape from Germany, having their meal in a separate room. A little later Kempe marched in, almost brushing Peter in passing, but of course didn't recognize him. Peter had to gloat inwardly, especially when his thoughts were centered on what he planned for the next days' sailing trip.

As agreed, in the afternoon Peter met Captain Volker in the park, where he told him of his new acquaintance the book store owner, and their little sailing excursion.

Volker already knew Kempe was guarded by the two *Gestapo* thugs, and was seriously concerned how he and Peter could prevent the U-Boat G-17 from leaving. Peter assured him he would do his utmost to foil Kempe's escape. Unobserved by anyone, Volker slipped a package wrapped in an old rag into Peter's hands.

"Here is the weapon with ammunition you asked for."

Peter thanked his friend, promised to make contact once he had completed his planned action, and left to wait until darkness. He returned to his haystack catching a few hours of sleep before the following days' events.

Around two o'clock in the morning, he returned to the farm, entering the back yard through a narrow gate. Under a partly clouded sky with the moon shedding a dim light over the scenery, Peter carefully approached the building and living quarters. A side door to the home provided entry from the farmyard.

The door was unlocked allowing him silently to slip into the hallway leading to various bedrooms. He knew the farmer and his wife slept in the room next to the kitchen, located at the far end of the corridor; he also knew that Kempe's room was next to the farmers.

Pressing his ear to the first door, he heard the sound of someone breathing heavily in deep sleep. Carefully, he eased the door open. The old handle lock with its black square shaped steel encasing operated quietly without a sound. Tiptoeing, Peter approached the slumbering *Gestapo* agent. On top of the night table lay the agent's pistol, a Walter model that Peter immediately pocketed, then putting his *Mauser* pistol under the man's chin he nudged him until he woke up.

"Don't make a sound or you are a dead man." He said roughly. "I want you to quietly get up, grab your trousers and jacket, and then we will both leave. I don't intend to harm you; I only want you out of the house. Do you understand?"

The agent was terrified, and his courage failed him. He seemed to manage control only when beating defenseless victims. Following Peter's orders, he walked ahead in his long white drawers, holding his clothes and shoes, heading towards the farm yard exit door. They moved silently across the yard, approaching the meadow leading down to the stream. Peter found an apple tree where he told the agent to get dressed, then, taking the man's belt, ordered him to lie on his back, straddling the tree trunk with his legs. He then tied the man's feet together at the ankles, making it impossible for him to get free; he couldn't reach his ankles because the tree trunk obstructed any such movement.

Peter again told the man he wouldn't harm him as long as he cooperated and remained quiet, otherwise he would experience the alternative. Satisfied, Peter returned to the building the same way he initially had entered.

Walking to the second door in the hallway he again pressed his ear against the door. The man inside was snoring noisily, making it easy to approach. His *Walter* pistol was also on the night table so Peter again slipped it into his pocket. The man awoke suddenly after Peter jabbed his *Mauser* under his chin, but before he could yell in alarm, Peter pressed his hand over his mouth and warned him to be quiet or he would have to kill him. Shaking with fear, he became silent and did as he was told.

Following the same procedure as before, they arrived at the apple tree orchard where Peter produced a piece of paper on which he had printed. "*Herr Standarten Führer*, knowing you are leaving us tomorrow, we have also decided to disappear.

We'll obtain new identities and try to continue our *Führer's* teachings. Wishing you and your comrades much success. *Heil Hitler.*"

Peter demanded that the men sign the document or he would have to shoot them. With shaky hands, both signed the letter and Peter returned to the building. He left the note on the night table in one of the empty bedrooms and headed again for the back exit. He had just closed the house door behind him to leave, when the farmer suddenly appeared in front of him with a pitchfork aimed at his chest. Just as the farmer lunged, Peter threw himself to one side, barely missing the fork as it crashed into the door. Peter drew his *Mauser*, fired and put a single round into the neck of the farmer, who fell, fatally wounded, to the ground.

Making sure the man was dead and before anyone would arrive to investigate, Peter ran swiftly through the yard's front entrance, up to the road, and crossed over to the other side, then heading into the small forest for cover. From there for a moment he watched, then raced over to the Mercedes, where the rear passenger door was unlocked. He crawled into the back, lay on the floor, and waited.

Ten minutes later the front door of the farmhouse opened and Kempe appeared together with the dead farmer's wailing wife, who had by now discovered the body of her husband. Peter heard Kempe, who had obviously found the note, cursing and threatening his treasonous *Gestapo* agents with severe punishment, blaming them for having murdered his farmer friend. He promised the woman he would drive to town and call for help, then jumped into his Mercedes, started the engine and headed toward Eckernförde.

In a quiet voice, pushing his *Mauser* against Kempe's neck, Peter said:

"Good morning *Herr Standarten Führer*, it's nice to see you again. Slow down or you will lose control of your beautiful automobile." Kempe froze in horror. Yes, it was Peter's voice, but how could it be? Peter was dead; he had personally drowned him!

Peter ordered Kempe to drive slowly toward the marine basin, leave the car and hand over the keys and his pistol. Jabbing his gun into the small of Kempe's back as they walked, Peter guided him to the little sailboat. He pushed him into the vessel and told him to sit at the front. Seizing an available line he tied Kempe's arms tightly against his body so he could barely move, and then roped him down to the seat.

By now it was about four in the morning, with only the moon illuminating the deserted surroundings. In the dim light, Peter found the key the old man had left under the plank. Releasing the chain holding the boat, he pushed off from the pier. The wind came from the southwest as he hoisted the main sail which took the boat gently out of the basin. It wasn't necessary to raise the jib, for the wind was favorable. They continued to move out of the harbor towards the open sea.

"I don't recognize you," Kempe said in a state of shock. "You're a stranger who has Peter's voice. Your disguise is incredible! Where have you come from? For heaven's sake, what are you trying to do to me? Maybe I was harsh with you, but as you surely must know, I had to follow orders. Please be reasonable. Let's talk. I'll do anything you want."

Peter was amazed at Kempe's inability to disguise his overwhelming fear. But he kept silent, steering the boat further and further away from shore. After about two hours, he turned the boat into the wind, lowered the mainsail and freed Kempe from his ropes, carefully pointing his pistol at him as he began to speak.

"Heiner, do you remember all these innocent, pitiful Jewish families you sent east to their death? Do you remember the young boy in Paris you executed like a piece of vermin? Do you remember the French hostages you ordered to be shot by *Schaarführer* Stolz? You are a ruthless murderer, and if Erika and I hadn't been such good swimmers you would have gotten away with our murders, too. Now, you have reached your moment of reckoning. I'll give you more than the chance you gave Erika and me, by offering you a swim back to shore. I'll count to ten, and by then you better have jumped overboard or I'll shoot you. I won't kill you; just wound you to make your swimming even more difficult. Maybe your blood will attract the odd shark."

Peter began to count. Kempe, his face drained of color, his body visibly shaking, began to beg for his life, but Peter continued counting. At the count of ten, Peter released a shot, barely missing Kempe's ear. Kempe jumped immediately over the side, crying for mercy. Peter turned the boat away from him and fired a couple of well-placed shots into the water. Hoisting the main sail, he brought the boat about and headed toward shore, with the moon pointing him in the right direction. Kempe's shouting and begging slowly faded into the distance. Peter, tying down the helm, raised the jib, allowing for a strong port tack in the desired marina direction.

At about six thirty in the morning, Peter heard the sound of strong diesel engines approaching from the west. In the dawns dim light a submarine headed toward him above water. Peter steered his boat to starboard to avoid a collision course, and the vessel passed by. To his dismay he identified the markings on the conning tower: "G-17". Undoubtedly the Captain could no longer wait for Kempe. Using his own Initiative, he had left Eckernförde, accompanied by his six Nazi companions for the voyage to South America. Hopefully, Peter thought, the British navy would discover the U-Boat and dispose of her and its criminal passengers.

In the meantime, Kempe desperately clung to his life, thinking, "How could Mertens have escaped from the submarine? How can I survive this freezing water?"

Soon he felt his jack-boots dragging him down. Swallowing huge amounts of salt water, choking, and with excruciating effort he succeeded in getting rid of the left boot, but the right one would not come off. The icy water was numbing his body through his uniform and only constant swimming kept him alive. Finally, Kempe managed to remove his right boot and propelled himself toward the shore in a westerly direction with powerful breaststrokes.

Suddenly, he heard the sound of diesel engines approaching from the west. Frantically swimming toward the heavenly sound, he realized a submarine was approaching. Immediately he started screaming at the top of his lungs for help.

Gerhard Lange, the new Captain stood in the conning tower, unaware of Kempe floating ahead of him. He pointed his binoculars toward the horizon. He wanted to be certain to keep his vessel in the channel, taking him and his passengers out of the shallows, on up north and around west into the North Sea. The diesels and open access to the below control room drowned out Kempe's frantic calls.

Realizing he was not being seen or heard, Kempe swam frantically toward the U-Boat, hoping to be able to grab a loose end or protrusion of the vessel as it passed. He knew it was his only chance.

The bow wave rolled him over and his body bounced along and against the side of the hull. Feverishly grasping and clawing at the passing steel surface, he succeeded only in slipping and slithering over the hull. Within seconds he reached the stem section of the vessel where the powerful surge suction from the propeller drew him down.

In panic, but too late, he realized the horror awaiting him as the gluttonous propeller instantly pulled him down into its churning vortex. Feet first, the huge volume of icy water sucked the screaming Kempe into the merciless guillotine-like blades. In an unstoppable revolution, his legs got caught and severed.

Open bone and flesh, discharged a red foaming bloody blotch onto the surface and into the silence of the Baltic Sea. He no longer was able to swim and slowly sank into the deep of the salty water, until he inhaled enough water in to his lungs to lose consciousness.

The Captain in his conning tower felt a thudding vibration and thought maybe a small piece of rubbish might have passed through his propeller, but the boat seemed unharmed, so he continued his steady cruising speed toward his final destination.

****

Sailing the small boat back and arriving at the basin's mooring, Peter reattached the security chain and placed the key under the plank. It was eight thirty and daylight.

Unlocking the Mercedes, he drove in the direction of the farm where, enroute, he found a small lumber trail leading into the forest. He left the car there and walked back to where he had hidden his bicycle.

Hoping to contact Hilmar Volker, Peter cycled back to Eckernförde.

Once again he sat on the park bench across from the boarding house. While waiting, he removed his uncomfortable make-up The ugly nose, bushy eyebrows and beard had done their job well. It was a painful task, especially removing the beard with its masking glue sticking to his skin.

Volker did not arrive. Tired of waiting, Peter decided to give the boarding house caretaker a letter for the Captain. He wrote on the hotel stationary that he had accomplished the main task, and signed it: "Your revived friend". Now he headed his bicycle in the direction of Flensburg and his Danish friends, where he hoped to arrive sometime during the evening.

Back at the farm, the dead farmer's wife had become impatient after nobody arrived to investigate her husband's murder. She took her bicycle, rode to the town's police station, and reported her terrible story.

When the police arrived at the farm to investigate the murder scene, they searched the property, and discovered the two Gestapo agents still tied to the apple trees.

After being released, the two agents reported that a sailor had attacked them, but they couldn't come up with a motive, especially since he ·didn't harm them. They wisely omitted the fact that they signed Peter's note, that Kempe had received, telling about their defection. It became apparent that the farmer had tried to skewer the intruder with his pitchfork, because the door panel showed deep indentations in the shape of five tines over the door's surface.

The two policemen asked if Kempe was still in the area. The farmer's wife reported that the last time she saw him was when he was driving his Mercedes toward town to call for help. The agents thought that Kempe might visit the submarine base, because today was his departure date. The policemen promised to look into the matter, saying that the first thing they had to do was find his Mercedes.

Toward evening Peter arrived at the home of his Flensburg friends, where he told them briefly what had happened at the farm, and his sailing excursion with Kempe. Afterwards, he was grateful that the Pedersen's gave him the opportunity to take a bath and remove the itching remains of his make-up.

CHAPTER 24

In early April, 1945, the British army arrived in Lübeck, eventually moving north to occupy Kiel and the province of Schleswig Holstein. Private homes, which previously had displayed Hitler's Swastika flags. now displayed white sheets of surrender from windows and balconies. Hitler's war was finally coming to an end.

Mertens decided to make contact with the first Allied troops he came across by cycling south towards Kiel. On the outskirts of Eckernförde, he saw a detachment of armored vehicles and tanks, together with a company of infantry soldiers in the distance. Slowly he cycled toward the group and dismounted, waiting for them to move in his direction. Four British soldiers with rifles at the ready approached him. Smiling, Mertens called them: "It took you a long time to get here, may I speak to your commanding officer?"

The soldiers were in no mood for joking. They were fully aware that their lives were still in danger, and knew that moving into enemy territory was serious business. Eventually, one of the soldiers shouted at him to shut up and raise your arms immediately. Mertens did as he was told, and one of the soldiers came up and frisked him for weapons, which he had wisely left in Flensburg. He had a hard time understanding the soldier's Yorkshire accent, but eventually came to understand that he wanted him to move further south on his own and make contact with the main body of the advancing troops.

Fortunately, a group of troop-carrying trucks started to pass. Two officers and their driver were in one of the leading jeeps. Mertens raised his hand for them to slow down or stop, which to his surprise they did.

"Gentlemen, I'm Captain Peter Mertens of the US-Army Rangers, I'm an American agent assigned behind enemy lines. I would like to speak to someone in authority."

The two officers spoke with each other, then ordered two soldiers to take Mertens further south to a recently established command post on the outskirts of

Kiel. Both soldiers rode motorcycles, so he was told to leave his bike behind and join them on one of the pillion-seats. Mertens threw his bike off the road, hiding it in some bushes, hoping he'd find it later.

The command post was located in a large farmhouse guarded by a number of red- capped military police soldiers. The motorcyclists left him there with a Sergeant of a British Intelligence Unit, who asked him to be seated in front of a makeshift desk. Mertens quoted his assignment's code name, "Blondie, " and asked the Sergeant to check with his superiors. He suggested they should contact Karl Banzer from the OSS, who also worked for the British Intelligence Service, and whom he had seen last in Strasbourg.

The Sergeant excused himself, calling one of the military policemen to stay with Mertens. About half an hour later the Sergeant returned and asked that Mertens follow him to meet Colonel Richard Appleton, who was in charge of the outpost.

The Colonel rose from his chair and extended his hand as soon as he saw Peter.

"Good to meet you, Captain," he said. "Welcome back to the winning side. But most of all, congratulations on getting back safely. I have just checked your identity and am pleased to say Karl Banzer is now attached to my group; in fact, he is at the moment on his way to join us. In addition, another group of your American intelligence officers, headed by Colonel Donald O'Connell of the OSS, will arrive shortly. All of these comrades you should be fully familiar with.

"Our main task in this area will be to round up German scientists, before they fall into Russian hands. We need your help on that project. In the meantime, my Sergeant will find quarters where you can wait until your people arrive and debrief you. Just to be sure, tell me the answer to our code 'Blondie'!" Mertens answered: "Jean Harlow!" The Colonel was satisfied. Mertens thanked him for his cooperation, and asked if he could arrange to have his bicycle recovered, as he had borrowed it from good friends in Flensburg. The Colonel promised to have the Sergeant take care of this matter. In less than an hour, after having given directions to the Sergeant, Mertens had his bike back.

The Colonel provided him with military identification, stating that he was an American Captain attached to his staff. In addition he supplied a plain uniform without markings, together with GI issue clothes and boots. Now Mertens could return the sailor disguise to his friends in Flensburg.

The Sergeant found him a room in a hotel at the edge of town, near the Colonel's farmhouse headquarters. The entire hotel had been taken over by British authorities, and reserved for the soon to arrive American and British task force members. Mertens was now free to roam the area, to search for the people who had helped him, and maybe who now needed his protection?

He knew he had to go immediately and search for his friend Hilmar Volker in Eckernförde. British Forces had just occupied the town. He' began by making inquiries at the boarding house, where he asked a middle-aged woman for the Captain's whereabouts. He gave her his name, and she asked him to return in one hour, when she hoped to have heard from the Captain. Mertens took his bike and headed for the city, where he located his sailing friend Walter Hellwig's bookstore.

The store was open. Mertens spoke to Mrs. Hellwig, who was very frightened by another soldier in uniform entering her store. He introduced himself and asked to see her husband, at which point she burst into tears saying: "Your British authorities have taken my husband away. charging him with hiding two *Gestapo* agents in our apartment over the store," she said.

"They don't believe my husband's explanation that the two men demanded to be hidden or they would shoot us both. A neighbor across the street saw the two agents enter our store and reported it immediately to the British. What shall I do? What will they do to him? Walter never was a Nazi."

"Do you know where they have taken him or what has happened to the two *Gestapo* agents?" Mertens asked.

"I believe he's in jail in the local police station, together with the *Gestapo* men. They were all taken in my home by British soldiers who came into our store and arrested everyone."

Mertens smiled, and told the distraught woman to calm down; he'd go immediately to the police station and get her husband released. The woman shook her head in disbelief, wondering how that could be done, after having seen the stern, unyielding attitude of the British authorities on their previous visit. She also wondered how a simple soldier without rank could have any influence.

The police station was now under control of the occupation forces. The German policemen had to obey the victor's orders. Mertens asked to speak to the officer in charge, and was promptly taken to a Lieutenant of the military police, where he introduced and identified himself. The Lieutenant called Colonel

Appleton wanted to verify Mertens' identity, and then asked him what he could do for him.

"I understand you are holding a Walter Hellwig, together with two *Gestapo* agents at this station. I would like to speak to Herr Hellwig, if that is possible? But first tell me what is he charged with?"

The Lieutenant confirmed the prisoners were at the station, and ordered Hellwig brought from his cell. He explained that Hellwig was found hiding two *Gestapo* agents. Mertens replied that he knew Hellwig, and that the man was not a Nazi, but was forced to hide the two agents or they would have killed him and his wife.

The two German policemen who escorted Hellwig into the interrogation room were astonished to see Mertens. Not long ago, they had held him in their cells, and then had him interrogated by Kempe. He was supposed to be drowned, together with this poor woman. They just could not make any sense from this most startling situation. Mertens was amused by their consternation, but ordered them to leave him alone with Hellwig. He greeted Hellwig with a smile, shook his hand and asked him to sit down.

"How did you get yourself into this ugly mess Herr Hellwig?" he asked in German.

Hellwig looked hard at Mertens. He rubbed his eyes and looked hard again, but still couldn't understand why this stranger, in plain English uniform, was so friendly to him. He had never met this man. 'What is going on, I do not understand!"

Mertens suddenly realized that Hellwig had only met him as a sailor, and even then in an ugly disguise.

"*Herr* Hellwig don't you remember our sailing excursion a few days ago?

I recall your generosity in lending me your boat. You have to forgive me, but I was involved in a very difficult assignment and needed a sailor's disguise. Actually, I'm an American officer who was assigned to work behind German lines. I know you're innocent of the charges brought against you. I'm sure my British colleagues will set you free, immediately."

What Mertens said took a while to sink in. Shocked but relieved, Hellwig expressed his gratitude.

"Hah, I thought so," he said. "I knew you were something out of the ordinary.

Your whole bearing and attitude during our boat trip told me you were not a typical German. I'm certainly glad we met and thank you very much for speaking on my behalf."

The British MP Lieutenant ordered the two German policemen back to gather Hellwig's belongings and release him immediately to his family. But just when they left, Mertens requested that the two *Gestapo* thugs be brought from their cells.

Within minutes the two *Gestapo* men stood in front of Mertens and the Lieutenant. The two German policemen had to guard both prisoners. The two Nazis looked utterly amazed when they recognized Mertens.

"Wasn't this American spy to be dead and drowned with his mistress? They just recently had jailed him and had witnessed his and Erika's drowning. How can he be still alive?"

Now the roles were reversed; and their former prisoner was in charge of interrogating them.

"Sit down! Do you know who I am?" Mertens asked.

"Yes, you are an American spy discovered by Dr. Weber and *Standarten Führer* Kempe!" they answered immediately.

"Yes, you're correct, but don't you remember the incident at the farm where I had the pleasure of tying you to the apple trees? I was quite lenient with you at that time; not knowing what kind of crimes you could possibly have committed. Unfortunately, I had to kill the Nazi farmer, but it was either his life or mine. My colleagues at British Intelligence will now begin to investigate your past. If you've committed no vicious crimes, you probably will have nothing to worry about."

Mertens had nothing further to add, so the Lieutenant sent the prisoners back to their cells, and asked him to tell whatever he knew about these two agents.

After Mertens related all the ordeals he suffered and problems he faced with the two men, he asked the Lieutenant for a reliable German policeman who might help him locate Dr. Weber, the Nazi fanatic.

The Lieutenant immediately found an elderly policeman named Wachtmeister (Senior Police Master Sergeant) Max Brinkmann. Mertens thanked the Lieutenant for his cooperation and left the station with Brinkmann.

The *Wachtmeister* also had a bicycle, so both men rode back to the boarding house, where Mertens hoped to find his friend Volker. Brinkmann was somewhat portly and a solid citizen with a wife and two teen-aged daughters. He was an ardent German patriot, but not an active Nazi, having never joined the party, even though they had a number of times requested his enrollment.

Now he was interested only in establishing law and order in his home area and obeying the laws of the new authorities. Recently he heard about the horrors of the Nazi concentration camps and SS atrocities. All this news made him severely depressed and greatly ashamed of his beloved fatherland. He always knew the Nazis did cruel things in these camps, but he could never have believed they were capable of devising and inflicting such total evil.

Volker came home at last. He picked up Mertens note from the caretaker. Now he noticed Peter and the policeman sitting in the lobby. Volker still wore his uniform. He planned to eventually surrender himself to the allied authorities.

Mertens explained his latest activities, and suggested that Volker join him by visiting British headquarters. At the same time, Mertens instructed Brinkmann to search for Dr. Weber, have him arrested, and put him in jail. Once this has been accomplished, he should report to him in person at British headquarters.

Riding their bicycles, both Mertens and Volker arrived at headquarters, where they were eventually shown to Colonel Appleton's office. The Colonel was astonished to meet a German officer in full uniform, including the *Ritterkreuz* around his neck.

After they sat down, Mertens explained that Captain Volker had saved his and Erika's lives. and that he was highly knowledgeable regarding the latest development of German secret weapons. Appleton was most interested in this information and pleased to meet the Captain; but he suggested that Volker might change out of his uniform into civilian attire.

With the war appearing to be almost over, Appleton asked Volker if he would work for him as an interpreter, helping to locate German scientists before the Russians could use their technical expertise.

Volker thought this offer over carefully. His first reaction was to refuse to cooperate with Germany's enemies; he had no wish to be labeled a traitor. But he also realized that this would be a great opportunity to fight against Communism and the Soviet Union. Finally, he decided to accept the offer.

The Colonel arranged that Volker could work together with Mertens providing daily 6 p.m. reports of their field activities, unless other matters made such meetings impossible. In that case, he'd appreciate being notified.

The new job offer persuaded Volker to stay at the boarding house where he kept all of his civilian clothing. He went there immediately to shed his naval uniform. After Volker had changed into a suit and winter topcoat, he and Mertens returned to Eckernförde, where they visited the police station and found *Wachtmeister* Brinkmann, jubilant because he just had locked Dr. Weber in a suitable jail cell. Mertens asked the Lieutenant in charge whether it would be possible for Weber to be brought up from the cells so he could confront him.

Once again, the two German policemen escorted a prisoner to the interrogation room, where they seated him in front of Mertens, the Lieutenant and Volker. "You obviously know who I am! Now what will you do with your Golden Party Membership Medal?" Mertens asked. Are you still proud to be a member of an organization whose goal it was to kill millions of innocent women, children, and harmless civilians with production line efficiency? How do you feel about that?

The expression on Dr. Weber's face was that of an innocent child, wrongly accused of some trivial misdeed. His watery blue fish eyes were greatly enlarged by the magnification of his bifocal glasses, and after clearing his throat a few times, he said, "I am a proud citizen of Germany and what you have said about killing millions of innocent women and children is obviously a Plutocratic Zionist lie," he said in a whimper of a voice.

"No German is capable of doing such terrible things to people, and certainly not our great *Führer*. I both resent and refute your accusation." Mertens knew how these Nazi fanatics thought and made a quick decision. He asked the Lieutenant to put Weber on a truck, under guard, and take him to the concentration camp at Bergen-Belsen immediately. The recently liberated place of horror was only three to four hours drive away. "Then after he toured that camp, bring him back here to continue the interrogation the next day at the same time," Mertens requested.

The Lieutenant thought the idea was excellent and promised to take immediate action.

Leaving Volker at Eckernförde, Mertens returned to headquarters to meet with Colonel Appleton. There, to his amazement, he met a whole group of American officers and civilians who had just arrived. He quickly found his superior Colonel Donald O'Connell, and his friend, Karl Banzer.

Later, having returned Dr. Weber entered the interrogation room with his head drooped forward and with none of his previous arrogance.

"What did you think about when you saw Bergen-Belsen?" Mertens asked.

'What I have seen has destroyed all my faith and belief in Adolf Hitler, including my former party comrades. When I spoke to some of the starved and abused victims, I realized how stupid I've been. I don't understand how I could have been so blind. All I can say is that nobody could believe we were capable of such hideous crimes. Believe me nobody in Germany knew such horrible things were going on in the concentration camps. Germany will have to pay dearly for this unthinkable outrage. I'm as guilty as any other supporter of the National Socialist Party."

"I'm satisfied with your statement," Mertens said. "The rest of your future is now in the hands of the law. Send him back to his cell."

Together with Volker they returned to headquarters where Mertens introduced him to his American friends.

Late at night Erika arrived per Jeep and was driven to Peter's hotel. She joined him in his hotel room. The reunion was a wonderful reward; after all these previous, hair-raising adventrues. The next day Peter introduced Erika to all his friends. She especially joyously celebrated her meeting with her father again.

At one important moment, Peter and Erika went to Dr. Rasche's room. Peter asked the scientist's permission to marry Erika. The two men embraced emotionally. "Welcome to my family," Rasche said. "I am proud to have such a great son in law."

*** *** ***

EPILOGUE

Fast British Navy frigates made sure no enemy could slip through the area of the English Channel between Dover and Calais. The weather, in early April, 1945, was partly cloudy with gusty southwesterly winds whipping up heavy waves and storm-driven rollers. Suddenly, the sonar operator on one of the frigates reported a submerged vessel passing on starboard.

The Captain and his officers on the bridge knew that only a U-Boat skipper with a death wish would risk entering this narrow straight under the watchful eyes of the British navy. With great confidence, the frigate Captain gave orders to shadow the passing submarine, approach it at a greater speed, and lay down a massive pattern of well placed depth charges in these relatively shallow waters. Within minutes, he reckoned, they should be precisely above the submerged target, and finish off this suicidal Nazi sub.

The operator reported from the sonar room that he couldn't track the target at its present speed. Exasperated, the frigate Captain signaled the engine room: "Full speed ahead". But the heavy head wind and waves crashing over the bow of the straining vessel made such a speed dangerous to sustain.

"How are we doing? Are we gaining?" The Captain called out impatiently.

"Negative, Captain," the sonar operator answered.

"The bloody U-Boat is increasing her distance by the minute. We cannot gain anything or even keep up with the contact!"

Furious, the Captain repeated his order on the bridge telegraph: "Give it all you have!"

But the Chief Engineer's answer from the engine room replied: "We're at our limit, we're reading almost thirty knots. If we keep this up in this weather, we'll fall apart!"

"Those bloody huns have concocted another of their damned new weapons," the Captain growled in a fury.

"Thank God the war is almost over; otherwise we'd probably have more U-Boat trouble to deal with!"

After the sonar man announced that he no longer could make contact with the U-Boat, the Captain gave orders to abandon the chase.

Did U-Boat G-17 reach South America? The answer is: "Very likely!" The fact remains that Professor Hellmuth Walter's Hydrogen Peroxide invention laid the foundation for one of our modern propulsion systems. The question presents itself: "Is it always necessary to start wars to achieve advanced technology? Today's fear of terrorism and the development of future superpowers seem to demand huge expenditures for more modern weapons. Can these weapons prevent war? Or will they ultimately result in the complete destruction of the human race?"